Soilborne Pathogens

Soilborne Pathogens

Urvashi Swami

Delve
PUBLISHING

www.delvepublishing.com

Soilborne Pathogens

Urvashi Swami

Delve Publishing
224 Shoreacres Road
Burlington, ON L7L 2H2
Canada
www.delvepublishing.com
Email: orders@arclereducation.com

ISBN: 978-1-77407-656-9 (Hardcover)

ABOUT THE AUTHOR

Dr. Urvashi is a microbiologist. She completed her Masters from Chaudhary Charan Singh Haryana Agricultural University, Hisar, after which she joined Panjab University, Chandigarh for completing doctorate. Her main area of research was fermentation and phytotherapy. After completion of PhD, she moved to Kuwait where she is running her own business of medical writing and editing.

TABLE OF CONTENTS

LIST OF FIGURES

LIST OF TABLES

PREFACE

Soil-borne pathogens provide the material basis of plant production. Conservation of the full range of these important soil resources is essential for sustained human progress and for man's continued development and utilization of natural resources. Conservation of soil systems biology is and will continue to be a vital issue for mankind. Loss of soil-borne biological diversity is an ecological hazard in that it heralds the loss of species valuable to the survival and development of mankind and could lead to genetic uniformity which breeds ecological vulnerability. Conservationists argue that reasons for conserving soil-borne pathogen diversities are economic or non-economic, prudential or ethical, utilitarian or non-utilitarian. There exists among writers and conservationists great diversity of positions with regard to the principal objective for conserving soil-borne pathogen diversities. To the ecologically minded conservationists, soil-borne pathogens represent an opportunity to advance human understanding of nature. Others emphasize the non-commercial utilization of at least some soil-borne pathogens believing that the principal objective of conservation is to provide pleasures of chase and conquest in farming. However, most writers use all these arguments in their attempt to gustily support soil-borne pathogenbiodiversity conservation.

Present book was compiled with utmost scrutiny of different journals, articles, research, and a wide array of soil-borne pathogens books. The main objective of preparing this book is to actually offer a base knowledge in the field of soil-borne pathogen regarding systems biology. This book is more informative to reach a wider public. It can be used by a wide range of readers who are directly or indirectly interested in soil pathogens, especially with respect tosystems biology. These include undergraduate and graduate students, pathologists, as well as governmental and non-governmental agencies working in pathological fields.

The book has 10 chapters, and each chapter discusses about a specific topic under the subject of soil-borne pathogens: Chapter 1 is anintroduction to soil-borne pathogen; chapter 2 is all about soil-borne viruses; chapter 3 discusses about soil bacteria; chapter 4 discusses about soil-borne fungi; chapter 5 presents information about soil-borne parasitic worms; chapter 6 discusses soil-borne protozoa; chapter 7 offers details about soil health management practices and how to sustain beneficial soil-borne pathogens; chapter 8 is all about how to control harmful soil-borne pathogens; chapter 9 discusses association between soil-borne pathogens and global climate change; chapter 10 discusses about the breeding for soil-borne pathogenic disease resistance.

These 10 chapters were comprehensively prepared for the readers to have a glimpse of systems biology of soil-borne pathogens. Great focus and emphasis were given to the simplicity of the content in order to make the book comprehensible and understandable to all readers.

Chapter 1 Introduction to Soil-Borne Pathogens

CONTENTS

1.1 INTRODUCTION

A pathogen is a microorganism which have ability to cause disease. The soil is the upper part of the earth in which plants grow. Therefore, soil-borne pathogens are microorganisms that live and survive in the soil that may cause disease in plants, especially in the roots. They may also affect the soil. Soil-borne pathogens could be Fungi, Bacteria, Nematodes and even Viruses. Like any other microorganisms, they require certain conditions to survive and thrive. These conditions could be environmental or physiochemical, such as temperature, pH, moisture, adjacent surrounding, etc.

Examples of some of these pathogens are Fusariumspp, Rhizoctoniaspp and Phytophthora spp. These are just but a few. They have different life cycles therefore one has to note the weather, soil conditions, location and other factors that would have a direct impact on the ill plant, such as an open end of a drainpipe or heavy foot traffic. These pathogens cause soil-borne diseases that may be characterized by rotting of roots, rotting of the stem, damping-off diseases, stunting, yellowing and spotting of lower leaves, wilt diseases, etc.

These diseases may cause many negative effects which includes reduced plant growth, increased costs to the growers and even damage to the environment. In view of this, we can agree that it is more beneficial to prevent than treat these diseases. There are many ways to control these diseases. To prevent a recurrence, one must eradicate the soil pathogen using methods such as crop rotation, thorough sanitation programs, pruning out dead crops, etc. Thereafter, one should prevent the development of the disease. This can be done by selecting resistant plant cultivars, choosing appropriate locations with proper drainage and sunlight, avoiding over-fertilization, etc. All these are ways to prevent soil-borne diseases thus preventing future problems, remembering it is always better to prevent than treat.

1.2 TYPES OF PATHOGENS

1.2.1 Viruses

An agent that is biological and multiplies inside the host's cell with life is termed as a virus. The hosting cell is compromised in giving out numerous of same looking entities of the previous virus at an imaginary level when infected by a virus. The cells that divide are not infected by the virus unlike most of the life processing organisms. The viruses that are new are gathered in the host cell that is contaminated. In contradiction to noncomplex contaminator agents, illnesses possess genes that provide the evolving ability to them so as to evolve thereby also mutate. Discoveries have been made for more than five thousand species of viruses (Katan, 2017).

It's not clear from where viruses emerged/originated from. Some have originated and evolved from plasmids (DNA pieces that are able to change positions amongst the bacterial cells), and in some circumstances, a given number may have grown out of bacteria. An infection (virus) is a component of more than one or three sections;

- Genes that are part of RNA or DNA
- Molecules (long) charged with the transport of information that is genetic
- A protective covering which provides the genes with security; and amongst other viruses, a covering containing lipid envelope which encloses the coat of protein and appliedtogether with receptors that are specific and required for successful establishment of infection.

Viruses differ in accordance with shape and structures. They also vary with respect to sizes. Viruses are passed from one host to another in numerous ways. Each virus species depends on a particular way for propagation, in a similar way there are specifications as to which tissue or the host to suppress by the viruses (Katan, 2017).

Viral infection can lead to illnesses in humans, animals and the plant species. They are in most cases done away by developing resistance against them. On viruses, antibiotics have no action. Vaccines are also used to prevent viral infections.

1.2.2 Bacteria

Literally, all animal life on earth relies on bacteria for them to survive since it is only bacteria contains the gene and enzymes that is useful to synthesize cobalamin and meet its demands through the food chain. Cobalamin is a vitamin that dissolves in water (water soluble) that is engaged in the metabolism of every cell of a human body. It is a cofactor in DNA synthesis, and amino acid and fatty acid metabolisms. It has an importance in the usual functioning of the nervous system through its function in the synthesis of myelin (Stirling et al, 2016).

There exist close to forty million bacterial cells in a small portion of soil and over one million in a millimeter of water. Bacteria are important in many steps of nutrient cycles by reusing nutrients such as the fixation of nitrogen. The nutrient cycle includes decay; bacteria have the responsibility for the putrefaction stage in the process. In biological communities, bacteria give the nutrients required to enable continuity of life by converting compounds that have dissolved like methane, sulfide and hydrogen, to energy (Stirling et al, 2016).

The largest number of bacteria is found in the gut flora, and on the skin. Many bacteria in the body are left with fewer possibilities to harm by the immune system, even though many bacteria have benefits, gut flora in particular. Several varieties of bacteria, however, are pathogenic and lead to diseases that are infectious to both the plants and animals.

1.2.3 Fungi

They are present in any habitat but mostly on land, many times in soil or plants rather than in the water bodies. The decomposer which is a group of fungi- grows in soils and others on nonliving matter of plants in which they perform a unique function of passing around other elements carbon included. A given section is parasite vegetation that leads to illnesses like scabs, rusts and mildews. In plants, diseases can contribute to loss of money for the farmer. Diseases in the animals are contributed by fungi that are few in terms of the number. In human beings, the diseases are like; athlete's foot, thrush and ringworm (Katan, 2017).

Types of fungi: Fungi are categorized in the account of their life cycles, availability or their structures and the type of distributional cells produced out.

The 3 known groups of fungi include:

- Microscopic yeast that possesses a single cell
- Filamentous molds that possess multiple cells
- Large fruiting forming bodies macroscopic filamentous molds

Yeasts are single cell organisms that are spherical in shape and are almost same sized as the red blood cells. Reproduction is by budding, where a new reproduced cell forms out of the original previous cell. Where buds have broken off on the yeast surface, scars can always be spotted. A species of yeast plays a fundamental function in bread manufacture and the process of brewing. Moreover yeast is nowadays used in cancer research also (Jenkins,

2005).

Macroscopic fungi with filament rise by giving out a mycelium from below the surface. They are different from the molds since they give rise to bodies that are visible that are spores holding. The spore-producing entity is composed of hyphae that divide and are tightly packed to give rise to separate fungal structure parts, like stem and caps.

1.2.4 Protozoa

Protozoa are single-celled eukaryotes. This shows that they are single-celled organisms that process nuclei together with a number of other useful organelles within the cytoplasm and enclosed by a membrane. They are in existence as a free-living organism or as a parasite. The different protozoa include; Paramecium, Giardia, Trichonympha, Trypanosoma and Plasmodium (Reddy, 2011).

Since they are eukaryotes, protozoa are large cells with a more complex structure. They have a DNA that is enclosed by a membrane, nucleoli, ribosome and multiple linear chromosomes.

Organelles in a cell vary from a cell to the other. There also exist a number of organelles that are exclusive of protozoa like:

- Contractile vacuoles
- The trichocysts of paramecium
- A given skeletal structure

Protozoa nucleus is vesicular. Due to this, the chromatic is scattered leading to a nucleus that is diffuse in appearance. Protozoa also possess locomotory structures such as cilia, pseudopodia and flagella that help in movement. These structures are also enclosed by a membrane.

Protozoa bring with them several hazards due to their diversity. They are known to be under the sub-kingdom Protista with very many species being considered as free-living. Free-living protozoa can be identified in almost every possible habitat. They have been grouped into six major phyla with majority of diseases leading to protozoa falling under the phyla Sacromastigophora and Apicomplexa (Reddy, 2011).

The free-living species of protozoa can be present in various habitats and particularly in water and soil. These types of protozoa have very little effect on human health due to the fact that they do not directly depend on other organisms for their survival. Some of the free-living protozoa can lead to pathology when introduced to the human beings' system. Others affect the human by means of releasing toxins into the body. Parasitic protozoa are the type that depends on the host for their daily living. Because of this, they live inside the host's body and may cause illnesses as well as other health complications.

1.2.5 Parasitic Worms

Parasitic worms (Helminthes)are big macroparasites. Protozoa are possible to be spotted by the eye without aid for clear vision. Most of them are worms of the intestines that are passed by soil thereby causes effect to the gastrointestinal tract. Others such as schistosomes reside in blood vessels. Some parasitic worms (leeches and monogeneans) are ectoparasites, therefore they don't qualify classifications as helminths, which are endoparasites. Parasitic worms find life and feed within living organisms. They are nourished and receive security while interfering with their host ability to take in nutrients. Parasitic worms cannot reproduce at all within humans. They have a life cycle that demands that they have to live outside humans at some stage. Helminths can live in the mammalian bodies for so long because of their possibility to copy the immune system of the host by producing immunomodulatory products. (Reddy, 2011).

All the parasitic worms produce eggs in their stage of reproduction. These eggs possess a strong shell that protects them from harsh environmental stress. These eggs can survive for a very long duration of time in the ecosystem; months or even years.

Many worms identified as helminths are the intestinal parasites. Infection caused by the helminths is known as helminthiasis or intestinal worm infections. All the helminths have a naming: the ending "-asis" (or in veterinary science: "-osis") is included at the end of the name of the warm to give the infection that particular warm name. For example, Ascaris is the name of a type of helminth, and ascariasis is the infection brought by the helminth (Reddy, 2011).

1.3 HARMFUL EFFECTS OF SOIL-BORNE PATHOGENS

Figure 1.1: A plant with rotting leaves that indicating signs of infection.

Sources: https://www.britannica.com/science/plant-disease/Symptoms

Harmful pathogens found in soil can trigger negative health effects to humans or plants. Some types of microbes can produce lethal toxins, which are fatal if the person inhales or ingests their spores, including transference through a wound. Some conditions that are likely to emerge include:

1.3.1 Crown Gall Syndrome

The pathogen Agrobacterium tumefaciens, is a type of bacteria thatcauses illness in plant tissue. In case A. tumefaciens infiltrates a healthy plant or tree via the stem or root from soil, it will immediately parasitize the host making it sickly and a potential carrier of the pathogen. Therefore, hosts of A.tumifaciens will succumb to growths and alterations in plant metabolism. Cancerous lumps may start as white callused matter on the plant. To inhibit crown gall sickness, it's necessary for crop life to be continued outside of polluted soil (Katan, 2017).

1.3.2 Anthrax

Anthrax is a potentially fatal human and animal diseased, which is triggered by the pathogen Bacillus anthracis found in soil. Generally, B. anthracis can persist for multiple years inside the soil, thereby making it quite dangerous as it can be spread across different generations. When the microbe produces spores, its potential for causing contamination becomes highthese spores can be easily dispersed during gardening. In fact, scientists are resolute that breathing of spores from polluted soil can indeed result in disease. Anthrax may further be responsible for generating a toxin which can result in fever, nausea, skin ulcers, and breathing problems (Katan, 2017).

1.3.3 Tetanus

Clostridium tetani is the germ that's responsible for triggering tetanus. C. tetani, generates spores which may be breathed. These spores can also infiltrate wounds, there causing tetanus in patients. If C. tetanienters human body, spore germination will happen almost immediately, thus allowing the microbe to grow. This bacterium produces a strong toxin called tetanospasmin. The substance affects the central nervous system, causing muscle stiffness, fever, spasms and respiratory difficulties.

1.3.4 Amoebic Dysentery

Entamoeba histolytica, a protozoan microbe, is known to cause amoebiasis or amoebic dysentery in people. They live primarily in tropical areas and are capable of burrowing through the gastric walls, spreading illness through the host's bloodstream to affect other organs, like the brain, liver and lungs. Amoebiasis is a disease of the gut instigated by an amoebae protozoan present in soils. It's responsible for causing intense diarrhea with blood, constant loose stools, spasms and abdominal pain when it enters the human system. In the environment, Entamoeba histolyticausually subsists in 2 forms which are, the free amoebae called 'trophozoites' and infective cysts. The latter isa set of amoebae encircled by a shielding wall, and they are passed through in contaminated soil or the carrier's waste. If humanconsume contaminated food which contains trophozoites, nothing much is likely to occur since they normally perish in the digestive tract due to its acidity. However, cysts are especially resistant to acid in the stomach, plus food polluted with cysts may pose an actual risk of dysentery infection. When these cysts enter the intestines, individual amoebae are discharged from them which leads to infection (Katan, 2017).

1.4 BENEFITS OF SOIL-BORNE PATHOGENS

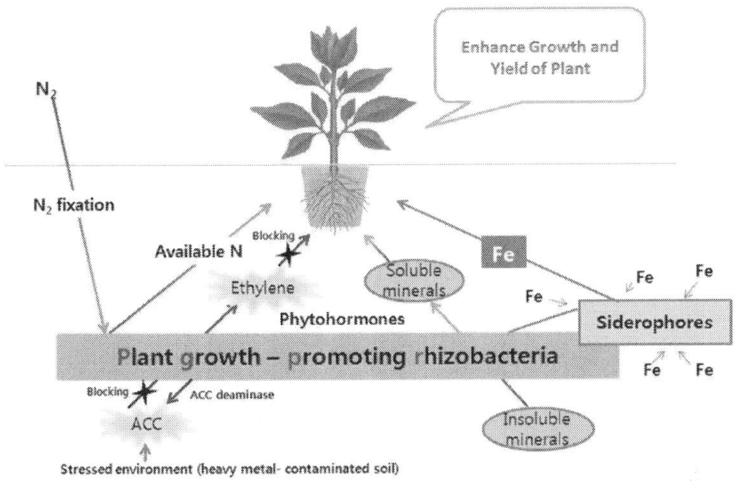

Figure 1.2: Soil-borne pathogens promote nutrient uptake and enrich the soil for healthy plant growth.

Source:https://www.researchgate.net/figure/Functions-of-plant-growth-promoting-rhizobacteria-Sources-http-wwwenvitopcokr_fig1_230636447

Pathogens grow in diverse micro-environments, including specific niches of the soil. Their populations expand quickly and the microorganisms are more active when easily consumable simple sugars are easily available within the rhizosphere. Simple sugars, complex polysaccharides, root exudates and dead plant material are quite plentiful in this part of the soil. Roughly 10-30% of all soil microbes found in this layer are actinomycetes, which are dependent on ecological conditions. A number of bacteria, for instance, generate a layer of glycoproteins or polysaccharides that cover the top layer of soil particles.

These compounds play a significant role in reinforcing sand, clay and silt particles to form stable micro aggregates that enhance soil structure. Useful germs live around the boundaries of ground mineral particles, particularly clay and other related organic deposits. They are vital in producing polysaccharides which fortify silt, sand and clay residues together, forming micro-aggregates and enhancing soil structure.

Usually, pathogens don't travel far in soil, meaning most of their movements are associated with locating water, developing roots or getting a ride on other soil-borne organisms like ants, earthworms and spiders (Katan, 2017).

Generally, most soil pathogens operate better in moderate pH soils that are well-oxygenated. Microbes provide huge amounts of nitrogen for plants which areoften missing in the soil. Several microbes discharge enzymes in the ground that catalyze phosphorus, to make it more solvable and plant accessible. Typically, germs tend to overpower fungi in cultivated or disrupted soils since the fungi favor more acidic ecosystems without soil disturbance.

Pathogens also dominate in watered fields since most fungi can't live without oxygen. Some can survive in arid or flooded environments due to their diminutive size, large numbers, and capacity to stay in small microsites inside the soil where ecological status may be favorable. When the environmental factors surrounding these microsites are more conducive, surviving microbes rapidly expand their populations. Likewise, within the soil, oxygen levels vary considerably from one environment to another. Huge pore spaces packed with air offer high amounts of oxygen, favoring aerobic conditions, whereas a short distance away, smaller micro-pores might be anaerobic or lacking oxygen. The variety in soil microsystems permits microbial life to flourish under different soil humidity and oxygen levels since even in saturated soils with limited oxygen, different kinds of microbes could live to replenish the soil when ecological conditions improve (Katan, 2017).

Natural procreation occurs in an array of plant ecosystems such as the soil. Microbes enhance the soil such that new crops can become established. In the absence of soil bacteria, plant populations and environments struggle to live or even exist. Germs alter the soil environment such that certain types of plants can persist and flourish.

Moreover, where fresh soil is forming, some photosynthetic microbes begin to populate the soil, recycling phosphorus, nitrogen and carbon, including other soil nutrients in order to generate the first organic material. A soil filled with good bacteria typically is cultivated or disrupted, plus has greater soil pH including nitrogen which is found in nitrate form. Pathogens make the soil's food-web to become more stable and diverse, ensuring soil nutrients are readily available within the environment and suited for plants.

1.5 HOW ARE SOIL-BORNE PATHOGENS TRANSMITTED?

1.5.1 Direct Transmission

Disease transmission of infection occurs when pathogens are carried externally or inwardly on the planting material or seed, similar to tubers, cuttings, sets and bulbs. Some common causes of fungal illnesses spread internally via seemingly healthy seeds are; Ring rot and Helminthosporin Blight syndrome. The bacteria responsible for causing them are transported inwardly through the host crop's tubers, and they usually affect potatoes. Similarly, the well-recognized red rot and whip smut diseases thataffect sugar-cane are fungal in nature, carried internally within the planting sets. The same applies to mosaic virus and leaf roll disease also in potatoes, which are transmitted through infected tubers.

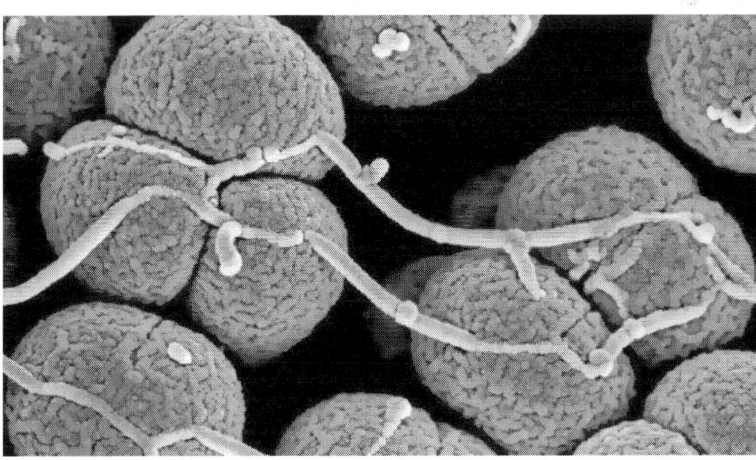

Figure 1.3: Soil-borne pathogens attacking a host cell and multiplying in number.

Source: https://microbiologysociety.org/publication/past-issues/soil/article/the-ecology-of-soil-borne-human-diseases.html

External transmission from seed or planting matter is the other common method. In this transmission mode, the microbe is transmitted externally across the seed's surface or vegetative broadcasted through plant parts such as tubers, bulbs and sets. Plus, it may further be transmitted as a physical blend of fungal components with the seed (Reddy, 2011). The typical grain smut is a good example of fungal structures that spread through the vegetative transmission. Whenever illness is established within a particular jurisdiction, pathogen transmission from one host to another or a place to the other is often termed as 'dispersal' or 'dissemination' of the microbial. Dissemination of crop diseases is distinguished in regard to various stages of infection.

In primary infection, pathogen contact is with an appropriate host plant, while instigation of the illness initially during the crop's season is known as 'primary infection'. Oftentimes, a few or multiple plants within the field may be primarily infected.

1.5.2 Indirect Transmission

Airborne: When pathogens are transferred as tiny particles in air-currents, it's known as airborne germ transmission. These infections diffuse when viruses or bacteria travel through dust particles, or tiny respiratory droplets which become aerosolized in case an infected individual sneezes or coughs. In whichever case, just about any airborne illness that's associated with microbes transmitted through the atmosphere can be categorized under this group.Airborne transmission of infection is dependent on various physical factors, which are endemic to the contagious particle. Some environmental factors that affect the efficiency of air-borne infection transmission are; temperature and relative moisture levels. The total sum of all attributes that influence humidity and heat levels, whether outdoor or indoor, including other factors influencing the diffusion of the droplets having the infectious particles, including winds and human behavior, contribute to the factors that influence the diffusion of airborne diseases. Furthermore, climate and environment also play an important role. Other factors such as rainfall, average daily sunshine hours, and latitude have a significant role.

Droplets: Droplet diffusion happens when microbes travel on somewhat large respiratory droplets, when infected individuals sneeze, exhale, cough or drip. These droplets travel just for short distances prior to settling, typically less than 3ft from the source. The droplets are laden with infectious particles which may be spread directly in case infected persons are near each other. Some soil- borne diseases that can be spread to humans this way includes; meningitis, common cold, influenza and mumps (Reddy, 2011).

Oftentimes, though, fomites might also be involved. Droplets can land on tables, hands, toys, mats and other similar surfaces, and can sometimes remain communicable for hours. Any hands that come into contact with these points, such as doorknobs, pens or telephones will also become contagious. If the contagious hand accidentally touches the eyes or nose, then the infection shall be able to pass into the new individual. For prevention, hand cleansing, particularly with disinfectant hand sanitizers, will help avert droplet transmission.

Foodborne: It's estimated that around 48 million U.S citizens get sick from what they consume every year. Around 128,000 further get hospitalized from pathogen infections. These food borne pathogens are responsible for diarrhea, abdominal cramps and vomiting among other illnesses. After consuming a foodborne disease there could be a short delay, or incubation period, just before the symptoms reappear. This delay may range from hours to days. During the incubation period, the microbes pass across the gastrointestinal tract and into the patient's gut system, fastening onto their cell wall lining, where they start to reproduce from there. There are certain forms of microbes which live in the intestine but aren't harmful. However, others like Vibrio cholerae, generate a toxin that makes the body to discharge water, causing diarrhea in the patient. Likewise, others, such as the typhoid bacillus, attack and replicate within the deeper tissues of the body. But not all foodborne microbes need an incubation time. It depends on the specific pathogen's characteristics.

Waterborne: Waterborne illnesses are caused by consuming contaminated or unclean water. This kind of water may cause various diarrheal diseases, such as cholera, including other serious conditions like Guinea worm syndrome, dysentery and typhoid. Water-based diseases are estimated to cause around 3.4 million fatalities each year, which is a major health concern, especially in developing countries. Illnesses can be transmitted while washing, bathing, or taking contaminated water. Whereas diarrhea and nausea are the most regularly reported signs of waterborne infections, other symptoms may include ear, skin, eye or respiratory problems (Reddy, 2011).

Microorganisms causing illnesses that usually are waterborne predominantly include; bacteria and protozoa and most of which are gastrointestinal parasites, or may attack the cells or cardiovascular system through linings of the digestive system. Many other waterborne disorders are triggered by viruses, but despite the differences associated with describing viruses, it's still practical and appropriate to simply consider them as microorganisms within this regard. Metazoan parasites form the single greatest form of waterborne pathogens.

Zootonic: Zootonic diseases, also referred to as zoonosis, are contagious illnesses caused by viruses, bacteria and parasites which disperse between vertebrates and humans. Some major modern infections like the Ebola virus, as well as salmonellosis, are categorized under zoonoses. Others are anthrax, dengue fever, bovine tuberculosis, hepatitis and lyme disease among others. Zoonoses may be transmitted through various ways such as; through the air, eating contaminated food, closeness with a diseased animal, insect bites such as mosquitos or ticks and handling a surface where a diseased animal also touched. Many times, transmissions happen when individuals go for outdoor activities like hiking, boating, or biking. Where they may get into contact with places where infected wild animals might have passed beforehand.

Petting zoos, also contain common places for a zoonotic disease to be transmitted. Plus, those who stay near livestock are at a greater risk of contracting zoonoses from ticks and fleas, which usually accompany most livestock (Reddy, 2011).

1.6 THE RELATIONSHIP BETWEEN SOIL-BORNE PATHOGENS AND HOSTS

In nature, plants exist as members of a diverse ecosystem that includes numerous and different microbes in the soil. It's long been held that some microbes, like the nitrogen-fixing symbiotic germ or mycorrhizal fungi, play vital roles in crop performance by enhancing mineral nutrition. Nevertheless, the whole range of microorganisms associated with crops and their capacity to substitute artificial farm inputs has only lately begun to be uncovered. Within the past few years, considerable progress has been achieved in understanding the arrangement of rhizospheric microbiomes including their dynamics. There is also clear evidence to show that plants determine microbiome structures, primarily by root exudates, but also that microbes have developed different adaptations to flourish in the rhizospheric ecosystem. These methods of interaction including the processes determining these changes in microbiomes are, nonetheless, largely unidentified and research is still ongoing.

1.6.1 Interconnection between Plants and Soil Microbes

While plant physiologists oftentimes regard soil as just a nutrients source for crops, it's really a complex environment that hosts fungi, bacteria, animals and protists. Plants showcase a wide array of useful relations with these soil-borne organisms, spanning the whole range of ecological options including; mutualistic, neutral, competitive, commensal and even exploitative in some cases. Throughout modern crop science, most interaction researches have focused on lessening pathogenic effects like herbivory and infection, even mitigating abiotic stress conditions. Nevertheless, there's also been enduring interest in describing the positive environmental interactions that enhance plant growth. Case in point, mycorrhizal fungi including the bacteria found in nodulated peas are considered to be root symbionts (Reddy, 2011).

Moreover, during the 1950s, farming seeds were occasionally coated with microbial cultures to enhance growth and produce. After 1980s, many divergent bacterial strains, mostly Pseudomonas and Azospirillum, were already discovered to be having crop growth-promoting effects.

1.6.2 Microbial Qualities and Bioavailability of Plant Nutrients

There are various theories put forward to expound on how pathogen activity may improve plant growth. First, it manipulates the enzyme signaling of plants or repels and outcompetes pathogenic microbial organisms. Some further increase the bioavailability of soil-based nutrients once introduced into the ground. Soil microbes are active in metabolizing recalcitrant types of soil-based nutrients to release these compounds for plant nutrition. For natural ecosystems, a majority of nutrients like nitrogen, phosphorous and sulfites are blended in organic particles and are thus minimally bio-accessible for crops. To access these components, crops rely on the development of soil-borne germs like bacteria or fungi. They possess the metabolic apparatus needed to catalyze and mineralize organic types of phosphorus, nitrogen and sulfites (Reddy, 2011).

The components found in these germ cells are subsequently freed, either through cell lysis, turnover or through protozoic predation. It liberates inorganic compound forms found in the soil, plus ionic species like phosphates, ammonium, sulfate and nitrates which are the favorite nutrient forms for crops. In natural environments, these microscopic nutrient alterations are crucial drivers of crop growth, and may occasionally form part of the rate-limiting phase in ecosystem productivity. In particular, the rhizospheric layer of soil is a hot-spot of biological richness along with this plant roots host a wide range of microbial diversity as well.

1.7 PATHOGEN GENOMES

Figure 1.4: Soil-borne pathogen genome causes devastation in a cultivated field.

Source: https://www.technologynetworks.com/genomics/news/soybean-pathogen-genome-assemble-324737

Pathogens may either have prokaryotic genomes or exist in the form of single-celled eukaryotic microbes and viruses. Typically, prokaryotic genomes are much simpler to sequence because of their smaller genome size, in comparison to eukaryotes. However, recently there's been improved efforts on sequencing Eukarya genomes, though extra efforts are still being made going into the future (Stirling et al, 2016).

1.7.1 Genome Analysis

Dynamic genomes with improved plasticity are essential to allow pathogens, particularly bacteria, to thrive in changing environs. With the support of increased material sequencing methods, it's possible for scientists to detect, compare and record most of these versatile genomic events. Keen interest is in comprehending how genomic factors lead to microbial development, including how these procedures may be interrupted for the prevention of genetic disruption.

Genomic diversity in microbes is caused by gene gain, loss, and rearrangement. Proper understanding and detection of different genomic dynamic factors are essential in the development of convenient therapeutic tools for combating pathogens (Stirling et al, 2016).

1.7.2 Pathogenic Genome Loss and Decay

Genome decay or genetic loss happens when a gene isnot used by the germ any longer, or in case an organism seeks to acclimatize to a new environmental niche. Sequencing strategies and microarray tests have shown a wide pool of pseudogenes present in various microbial pathogen species. For instance, Mycobacterium leprae, which is among the most prolific microbes has been discovered to contain roughly as manypseudogenes as its normal function genes. But it is not the only microorganism exhibiting such traits, similar properties have also been found in Yersinia pestis which is a plague pathogen, including Salmonella enterica.

1.7.3 How do Soil-borne Pathogens cause disease?

Plant pathogenic microbes are disseminated in various ways. Dispersal commonly occurs through wind-blown soil, including sand particles which cause crop wounding, especially during or immediately after rains or a storm. Crop wounding is vital for the entry of many pathogens into a plant.

Some plant syndromes also require specific temperature conditions to spread and cause disease. For instance, Pseudomonas syringae only causes illness in temperatures below 22°C, and Xanthomonas campestris in temperatures over 22°C on legumes. Also, diseases may occur concurrently under growth conditions whereby day-and-night temperature fluctuations permit for infection progression in susceptible crops.Contaminated seeds or any crop part may be a likely cause of bacterial inoculum. Though machinery, packing material, clothing and water may as well disseminate pathogens, including bugs and birds. The recurrent monoculture within an area may further encourage increases in inoculum, thereby allowing microbes to be disseminated (Stirling et al, 2016).

Host-pathogen exchanges: Infection of crops by pathogens may occur in several ways. Contamination is generally taken to be passive, that is, accidental, even though a few instances of crop chemo-attractants have also been reported. Microbes can be drawn into a plant via innate plant openings like hydathodes, lenticels or stomata. Similarly, they can infiltrate through scrapes or cuts on leaves, roots or stems through placement by particular feeding pests. The nutrient circumstances in crops may be so as to promote multiplication in various plant parts, such as roots or flowers.

Furthermore, wind-driven rain carrying pathogens may be a way of spreading illness borne microbes across fields. Bacteria are also commonly introduced into crops through aerosol-mimicking wind-borne rains, or even vacuum infiltration (Stirling et al, 2016).

Chapter 2

Soil-Borne Viruses

CONTENTS

2.1 INTRODUCTION

A soil-borne virus is from the Furovirus genus consisting of a rigid rod-shaped frame, and positive sense ribonucleic genomes having two strands packaged into distinct particles. These two strands differ in length, where the shorter one measures approximately 138-160 nm and is around 10 times more predominant in infected matter. Compared to the longer strand version which has an average size of 281-300 nm. However, both particles are needed for an infection to be categorized.

Figure 2.1: A plant leaf shows signs of wilting caused by the soil-borne virus.

Source: https://www.agric.wa.gov.au/lupins/lupin-foliar-diseases-diagnosis-and-management

2.1.1 Symptoms of Infection

The first visual sign of soil-borne virus is uneven, chlorotic blotches in the farm which suggests a viral infection, though similar factors could be triggered by various other diseases or even biotic and abiotic elements. In dry areas, spots of symptomatic soil-borne infected crops usually happen in low-ground, wet zones of the field which are prone to contagion by the floating zoospores of the disease's viral vector (Emerson, 2019).

Moreover, upon keener inspection, you'll find that infected plants have stunted growth plus the leaves could bear a distinctive chlorotic mosaic or uneven mottling and vein lining. Some few varieties of the virus are also known to trigger resetting, whereby leaves and tillers continue remaining short, plus growth is compacted and tillering quite excessive.

In most cases, symptoms for this disease appear before late spring when temperature levels are below 20 °C. It could even be misdiagnosed to be a nutritional issue by some farmers, who may seek the wrong treatment due to their poor understanding of the virus. Over the recent years, distribution of viruses has expanded to eastern and central parts of the U.S. as well as Europe (Emerson, 2019).

2.2 HOW ARE VIRUSES CLASSIFIED

2.2.1 Double-stranded DNA (dsDNA)

The double-stranded DNA virus is also known as dsDNA, it's a rare infection that mostly affects lower plant varieties like algae. These organisms are large-sized with genomes measuring from 160-560 kb and containing up to 600 protein-synthesizing genes, which makes them considerably different from infections that affect higher plants. dsDNA iscommonly found in watery environments all around the environment and play vital, yet largely unnoticed, roles in managing algal communities like the dissolution of huge algal blooms typically known as brown or red tides (Emerson, 2019).

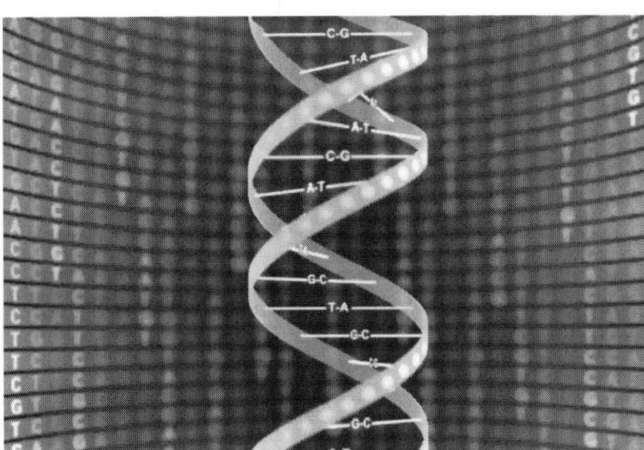

Figure 2.2: Structural makeup of a double-sided DNA, with a pair of arms circling each other.

Source: https://www.thoughtco.com/double-helix-373302

Replication method: Double-stranded DNA virus replication refers to the development of cellular viruses during the contagion process in the final host cells. Viruses should first enter into the cell prior to any viral replication.

According to virus's perspective, the role of replication is allowing for regeneration and future survival of the organism. By producing abundant duplicates of its genes and converting these copies into infectious viruses, the organism is able to persist in infecting new hosts. Typically, double-stranded viruses infiltrate the host nucleus first before they can start to replicate. Some viruses, however, will first require the target cell polymerases to duplicate their genome, whereas others, will encode their own unique multiplication factors. Whichever the case, duplication of the virus genome is largely dependent on a molecular state that's tolerant of DNA replication, hence on the cellular cycle. The organism may prompt the cell to compellingly undergo cell multiplication, which could lead to cell alteration and, eventually, growths or lumps. The most distinguishing factor about dsDNA is that they each have their own special ways of replication (Emerson, 2019).

2.2.2 Single-Stranded DNA (ssDNA)

There are 2 main families of ssDNA organisms known to affect plants, which are Geminiviridae and Nanoviridae. The former is the biggest known group of single-stranded DNA illnesses. It comprises a wide array of plant-based viruses such as; maize streak virus, bean golden-mosaic virus and beet coiled top virus among others. Together, these viral particlesare responsible for quite a large portion of crop damage globally.

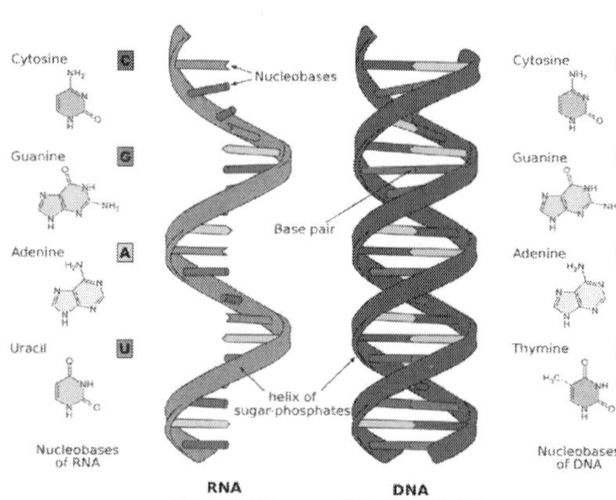

Figure 2.3: Comparison between the structural makeup of an RNA and DNA.

Source: https://www.thoughtco.com/dna-versus-rna-608191

The genome may either be an exclusive compound of around 2500-3100 nucleotides, but, for some single-stranded DNAs, it could develop as a set of similar-sized compounds measuring from 2600 to 2800 nucleotides (Emerson, 2018). Characteristics: These viruses have extended, geminate capsids and a pair of partially formed T=1 icosahedrons fused at the absent vertex. Also, the capsids vary in dimension from 18 to 20 nm in width and a length of around 30 nm. Viruses further possess two constituent genomes divided into two separate particles, plus both are mostly transmitted together for purposes of propagating a new illness within a target host cell.Just like most viruses, ssDNA genomes encode just a few proteins, meaning they are dependent on various target cell factors for duplication. Moreover, replication usually happens within the plant cell's nucleus through a rolling circle system. The resulting virus is arranged into germinate components in the nucleus (Emerson, 2018). However, it's not yet evident if these elements can then move from the nucleus, later being transmitted to adjacent cells in form of virions, and whether ssDNA is transported from one cell to another through the plasmodesmata. Viruses are often hosted in plants that lack enzymes for restriction DNA replication.

2.2.3 Reverse-Transcribing Viruses

The reverse-transcribing virus uses encoded enzymes to disrupt the genome of a host cell, thereby making it susceptible to diseases. Reverse transcriptase is vital to the contagious nature of retroviruses, most of which are responsible for diminishing the immunity of host cells. These viruses work by catalyzing the transcription of ribonucleic acid into the deoxyribonucleic acid, thus the name reverse-transcribing virus or retrovirus (Emerson, 2019).

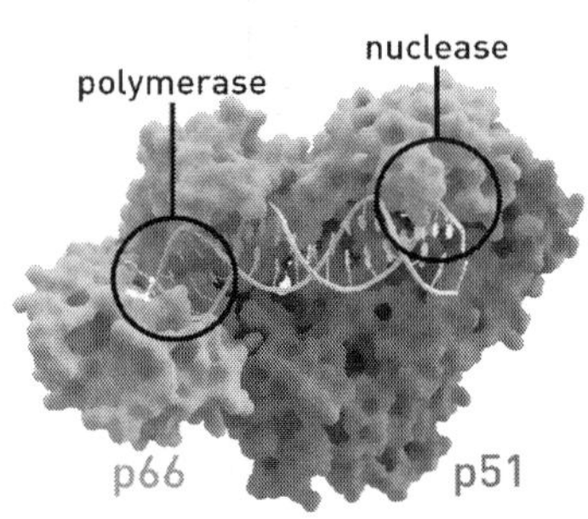

Figure 2.4: A reverse-transcribing virus with both the nuclease and polymerase parts.

Source: https://en.wikipedia.org/wiki/Reverse_transcriptase

Structural makeup: Viruses comprise an RNA genome covered within a protein casing which is protected by a lipid layer. Their genome features a total of 3 genes which are; the group-oriented gene, polymerase genome and envelope gene. The polymerase gene serves to encode the enzymes integrase, protease and reverse transcriptase, which propel the process for retroviral infection. When a retrovirus enters a host cell, in a process facilitated by protease, it usually overruns the host's genomic transcription machinery to develop a DNA provirus. The procedure, involving transformation of retroviral RNA into proviral DNA, is fast-tracked and mediated through reverse transcriptase.

Reversal mechanism: Retroviruses use a 'right hand' configuration that's similar to those seen in viral nucleicacid polymerases. Apart from the transcription role, they also have a component fitting the RNase H family, and this is crucial to their duplication. By degenerating the RNA template, these viruses permit the opposite filament of DNA to be blended. Moreover, some elements from the assimilation process act as primers for DNA polymerase, which is necessary for generating the second strand. Moreover, self-replicating bands of eukaryotic genes, also called retrotransposons apply reverse transcriptase processes to relocate from one point onto another point of genome through RNA intermediate processes (Emerson, 2019).

2.2.4 Double-stranded RNA (dsRNA)

Double-stranded RNA germs represent a wide group of viruses that vary immensely depending on the target host cell. Other factors that affect its functioning are; virion arrangement and genome segment ratio. This group comprises of rotaviruses and pathogens, plus recently there's been significant progress in discovering, at atomic and sub-nanometric categories, the makeup of various viral proteins as well as the virion capsids found in multiple dsRNA viruses. This has been helpful in highlighting the considerable differences in structure and duplicative roles of most of these viruses. By offering exclusive insights into basic aspects of structure-function interactions in virus atoms, particle assembly, connections, and pathogenesis, systems for the creation of original antiviral approaches and agents may be formulated.

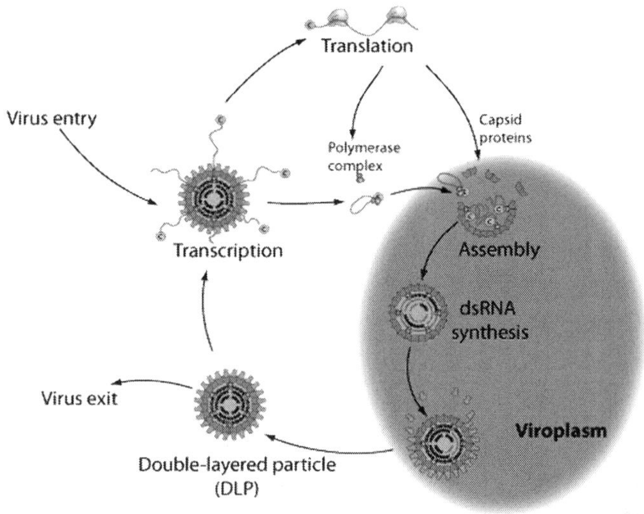

Figure 2.5: Replication process of a double-stranded RNA virus.

Source: https://viralzone.expasy.org/1936

Viral function in plant cells:Double-stranded RNA is a powerful trigger of genome silencing among plant structures. This silencing works through sequence-targeted RNA degeneration. For plants, a primarily common mode of silencing endogenous genomes is by altering the plant using gene construct programming. Targeted hairpin RNA consists of a reserved replication of a segment of the genome sequence, which is separated by a spacer that allows for easier plasmid construction.

Cell entry process: The dsRNA cell entry procedure is facilitated by a sequence of molecular linkages and readjustments. Moreover, some rotavirus entry methods like proteolytic priming, virion uncoating and initial attachment conform to well-grasped stages in the entry routes of other viruses, thereby promoting extrapolation from information specific to the dsRNA virus. Nevertheless, the most significant step in virus cell entry is the passage of the large non-enveloped component via a lipid bilayer (Gareth et al, 2018).

2.2.5 Negative Sense Single-Stranded RNA (-ssRNA)

The virus utilizes a negative sense, single-filament RNA as its genome material. They can be classified into positive or negative based on the RNA's form or polarity. The virus is also corresponding to the mRNA, plus it should be transformed into a +ve RNA by RNA-polymerase prior to translation.

As such, the filtered RNA of a -ve sense virus is not contagious by itself, since it should first be changed to a positive-sense RNA for duplication. These viruses are classified under Group V on popular germ catalogs like Baltimore classification. Additionally, negative-sense single-filament RNA organisms have multifaceted genomic sequences, cellular cycles and duplication habits that employ various protein constituents to attain specific conformations and perform various processes for sustenance and replication of their genomic chains.

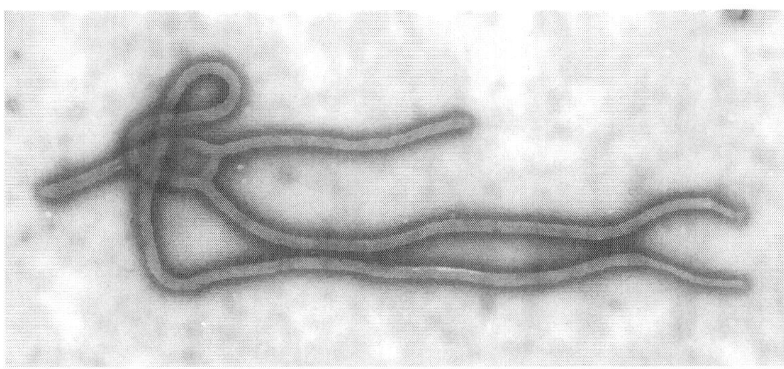

Figure 2.6: Microscopic representation of an -ssRNAvirus.

Source: https://en.wikipedia.org/wiki/Negative-sense_single-stranded_RNA_virus

The multifaceted structure of –ve sense single-filament RNA viruses is conveyed into their ability to subdue the innate immune reaction of the target cells it affects, including the capsid's construction, which is exclusive to the different categories of –ve sense single-filament RNA viruses.

Replication method:-ve sense ssRNA viruses require RNA polymerase for purposes of developing a +ve sense RNA. The positive-sense component functions like a viral mRNA that's converted into proteins, which is useful for the generation of new virion components. Likewise, once the virions are freshly formed, additional –vesense RNA particles are produced. In terms of measurement, the genome scope of a –veRNA virus goes between 10–30kb. Plus, there are two genome subtypes that can be identified under this type of viruses, which are segmented and non-segmented.

2.2.6 Positive Sense Single-Stranded RNA (+ssRNA)

The positive-strand RNA virus consists of the single biggest category of RNA viruses, having a total of 30 families. These viruses can be categorized based on their structure or polarity of the RNA to form both a negative and positive sense, including antisense RNA viruses.

Generally, positive-sense virus RNA is comparable to mRNA, thus it can be instantly decoded by the target cell. Negative-sense virus RNA mostly serves a complementary role to mRNA, plus it should first be changed to positive-sense RNA through the polymerase process before conversion (Emerson, 2019).

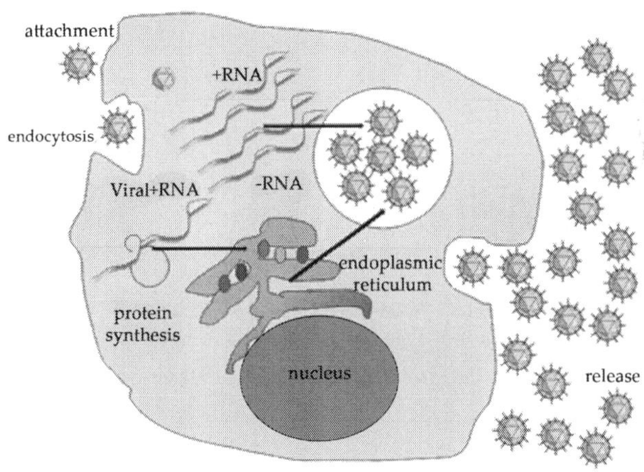

Figure 2.7: +ssRNA virus affecting a host cell after entry.

Source: https://courses.lumenlearning.com/boundless-microbiology/chapter/positive-strand-rna-viruses-in-animals/

Hence, purified RNA molecule of a positive-sense germ can instantly cause contagion, though it's still not as transmissible as the whole virus particle. As for the purified negative-sense virus, it's usually not contagious on its own as it must first be transformed into positive-sense RNA. Every virion molecule can be programmed to multiple +ve sense RNAs. On the other hand, Antisense RNA viruses look like negative-sense RNA particles, only that they further translate genomes from the thread. A common viral positive-strand RNA virus that infects humans includesthe picornaviruses.

Structure:Smaller RNA viruses are usually non-encapsulated, positive-stranded molecules having an icosahedral capsid. Moreover, the RNA genome is atypical since it contains a protein element on the 5′ final part which is applied as a transcription primer by RNA polymerase. The viruses are further divided into a massive array of genera, which affect different plant species, plus they can be activated or destroyed depending on the soil's pH conditions. Viral populations don't replicate through cell division like other microorganisms since they're a cellular.

2.3 IMPORTANCE OF SOIL-BORNE VIRUSES

2.3.1 They Assist in Research

The study of soil-borne viruses has a lengthy history. As a matter of fact, it's in plants that the viral genome was first discovered. During the late 1850s, there was a destructive disease thatwas scattering across tobacco fields in the Netherlands. Researchers at the period discovered that, injecting fluid from infected crops into vibrant ones transmitted the symptoms of mottling and plant-leaf discoloration.

At first, scientists assumed that the condition was caused by a bacterium. But additional tests done in the 1890s revealed that the contagious agent dispersing the disease could go through the small pores of a ceramic water filter, which is way too small to permit the passage of all existing bacterium species. Hence, in 1898 the term "virus" was coined by a Dutch microbiologist, MartinusBeijerinck, to describe the mysterious infection, but it would be a few more decades before scientists pinpointed its exact nature (Gareth et al, 2018).

During the 1950s, researchers started applying electron microscopy as a way of viewing plant-virus connections in detail, showcasing large cellular re-arrangements in diseased cells. They noticed that there were various structures resembling vesicles. In healthy crop cells, there are chloroplasts, mitochondria and nucleus, but in virus-infected cells it's only novel organelles that exist.

2.3.2 They can help Scientists better understand Human-Related Illnesses

For more than two decades, crops have been studied for their abilities to function as production podiums for bio-pharmaceuticals, like vaccines and monoclonal antigens. Much recently, plant-based viruses have been developed as non-toxic nanoparticles which can target various cancer strains and empower the body's immunity to slow or further reverse tumor progression (Gareth et al, 2018).

2.4 ROLES OF SOIL-BORNE VIRUSES IN SOIL FERTILITY

Soil fertility consists of three correlated factors, which are physical vitality, chemical composition and biological fertility (Gareth et al, 2018). In the latter case, microorganisms that reside in the ground and mix together with other nutrients may considerably improve a soil's richness and complexity. Moreover, a better comprehension of soil-microbiology is vital if farm production should meet the requirements of an increasing world population. But in many regions, the microbial population is still under threat from environmental pollutants affecting agricultural practices.

Soil viruses are very important since they influence the framework of soil biological factors through an ability to transmit genes from one host to another. Accordingly, viruses are key players in international crop cycles, affecting the produce turnover by complementing amounts of beneficial nutrients and gases in the soil (Gareth et al, 2018).

Despite these factors, the topic of soil-virology is still not well studied. Scientists are only beginning to understand that soils potentially harbor multiple novel viral species which, together, could represent a massive reservoir of genome diversity. There are number of researchers who believe that studying this widely unexplored variety of soil viruses could potentially influence our understanding, touching on the role soil viruses play in international ecosystem processes. Including the development of microorganism life itself (Gareth et al, 2018).

Collectively, viruses play a central role in disintegrating and degrading organic matter, recycling nutrients and enriching the soil. Without this recycling of elements, the perpetuation of biological life on earth might be next to impossible. Some viruses also produce sticky substances, such as mucilage and polysaccharides, which help to improve soil aggregation and nutrient absorption rates.

2.5 VIRUSES TRANSMISSION FROM SOIL TO PLANTS

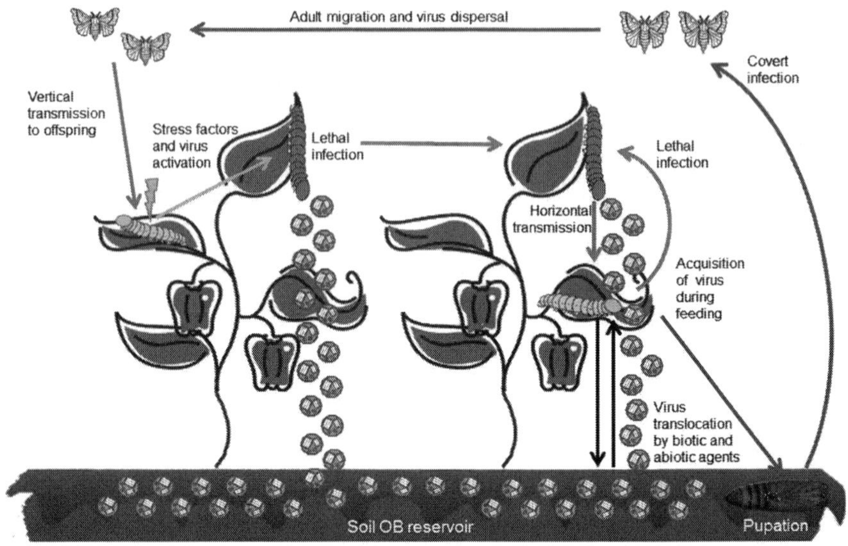

Figure 2.8: Vector insects transmit virus spores among different plants and the soil.

Source: https://www.frontiersin.org/articles/10.3389/fmicb.2017.01337/full

2.5.1 Horizontal Transmission

Horizontal transmission is considered as the spread of viruses between plants of similar species, but not the parent-plant and its offspring. Since the evolutionary destiny of the agent is not associated with propagative successes of the viral host, horizontal transmission often seems to progress virulence. It's therefore, a crucial aspect for evolutionary germ transmission studies. In this model of contamination, the plant-virus is usually carried from an external source. For purposes of "invading" the crop, the virus should first penetrate through the plant's exterior protective layer. Also, plants that have undergone damage caused by weather, trimming, or plant vectors are generally more prone to viruses. The transmission may also happen through certain artificial systems of vegetative replication, usually applied by farmers and gardeners. Such as grafting and plant cutting, which are widespread means by which crop viruses may be spread (Emerson, 2018).

Vector carriers: Sometimes, in biological transmissions, the plant virus may be carried through other species, like insects. Due to their highly structured social system and packed population density, honey bee colonies provide a rife environment for spreading infectious diseases in plants that they visit. Bee-carried viruses are usually transmitted via the proboscis which they use to draw nectar from plants, if these microorganisms are present in one plant then they can easily be transported to another, in case the bee lands on both of them.

Satellite viruses: Satellite viruses are contagious particles which complement the functions of horizontally-transmitted viruses, they usually cause plant infections by interfering with certain crop genome functions. These may cause symptoms like ringspots, leaf yellowing and deformed growth among others.

2.5.2 Vertical Transmission

During vertical transmission, the virus is transferred from a parent plant, and this may occur through any form of reproduction. Such as vegetative multiplication, whereby the offspring stems are innately similar to one main plant. For instance, when a new plant forms from the bulbs, roots or stems of the parent crop, the virus would be passed along with it to the maturing plant. In seed-based reproduction, viral dispersion occurs from seed infection.

Mostly, scientists haven't been able to get cures for crop viruses but instead, have been concentrating on decreasing the occurrence and spread of vertically transmitted viruses. Within the last few years, plant biologists have sought to recognize which factors control the progression of plant diseases and parasite virulence. Among the issues identified is the method of transmission, including the host plant's adaptation mechanism in regards to parasite evolution.In a study where seed vertical-transmission rate, viral accumulation, plant growth effect and virulence of both evolved and non-evolved germs in parent plants was measured. It was discovered that vertical passaging caused evolution of the virus towards improved vertical transmission mechanism, this was further associated with decline of virus build-up and virulence. Generally, the observed upsurge in seed transmission, virus accumulation reductions and virulence in vertically transmitted viruses wereinfluenced by reciprocal host adaptations for vertical passages, which consequently diminished virulence and reproduction of vertically transferred viruses (Emerson, 2018).

Indeed, variations in virulence have largely been tied to the effects of these organisms on host population trends, reduction of environmental biodiversity and occurrence and recurrence of infectious illnesses.

2.6 DISEASES IN PLANTS CAUSED BY SOIL-BORNE VIRUSES

2.6.1 Tobacco Mosaic Virus

Tobacco Mosaic Virus (TMV) is a plant-based virus that belongs to the Tobamovirusgenusring spots a rod-shaped particle measuring around 300 nm x 15 nm. Essentially, a single TMV atom comprises of 2,130 duplicates of the coatprotein (CP), which covers the RNA particle of roughly 6,400 nucleotides. The single-filament RNA encrypts 4 genes which include; a pair of replicas-associated proteins, transmission protein and a sub-genomic coat protein.

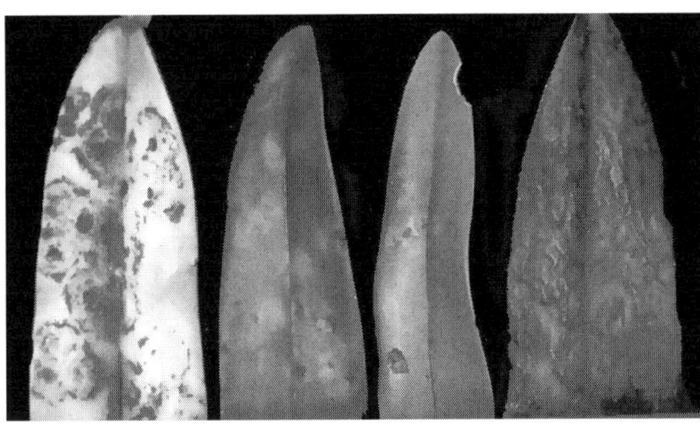

Figure 2.9: Tobacco Mosaic Virus mediated infection.

Source: https://www.forestryimages.org/browse/detail.cfm?imgnum=1525471

Classification: In terms of shape, it's a small, rounded ssRNA particle commonly found in wild tobacco plant varieties. The organism is also classified among helical viruses and is considered among the first viruses studied by scientists at a molecular level.

Even more, this RNA-comprising virus's capsid consists of roughly 2130 subunits of one structural protein containing157 amino acids, having a molecular weight of 17,400. The 2130 proteins are packed to a span or distance of 3000 angstroms, surrounding the axis and enfold a cylindrical space having a width of 40 angstroms.This space is useful for containing the RNA genome. Additionally,, viruses with helical structures such as this one, may differ in terms of the numberof subunits they utilize and the average number of proteins found in capsids. Also, more complicated helical viruses have been known to be enfolded in lipid membranes (Emerson, 2018).

Transmission from one plant to another:The TMV virus is easily transferred when a diseased leaf brushes against a healthy plant's leaf, through contaminated tools, and sometimes by farm laborers whose hands get contaminated with the germ after smoking tobacco cigarettes. Wounded plant cells also provide an entry point for TMV.

2.6.2 Barley Yellow Dwarf

Barley yellow dwarf syndrome was first discovered by Oswald in 1951, with further findings made by Houston in 1953. The disease is triggered by a genus of luteoviruses called barley-yellow dwarf luteoviruses. Some of the disease's strains and primary vectors include; Sitobionavenae, R. maidis, Schizaphisgraminum, R. padi, S. avenae and Rhopalosiphumpadi among others.

Figure 2.10: Barley Yellow Dwarf mediated infection.

Source: http://postrockextension.blogspot.com/

This disease is caused by a phloem-limited virus which is commonly transferred by aphids. Moreover, due to massive harvest losses in farming, this virus is, particularly of significant economic relevance. Since the management of this virus by itself is not feasible, tolerant barley strains are regarded as the most effectual approach to prevent yield losses. Even though various genomes and quantitative feature loci are recognized and applied in barley development for virus tolerance, still little is evident about the molecular and biological circumstances of this trait (Emerson, 2019).

Symptoms of infection:The signs of Barley Yellow Dwarf are dependent on the target cell's strain and cultivar, age and physiological condition of the host crop during the period of infection. Other factors that influence its transmission rate are; the strain and ecosystem conditions which may sometimes be mistaken for nutritional and abiotic factors.Some common signs include leaf staining from the apex to bottom sections, including from the margin to center. This staining of leaves may take on multiple colors depending on the crop affected. In barley, leaves usually become bright yellow; while in oat, the discoloration may be seen as orange, purple or red while in rye, wheat and triticale, the diseased leaves are normally yellow and occasionally even red.

2.6.3 Sugarcane Mosaic Virus

Also known as the mosaic of abaca, the Sugarcane Mosaic Virus was first discovered in 1919 by a scientist called Brandes. Some strains of this infection are known to infect specific host sugarcane species and are referred to as abaca mosaic-potyvirus.

Until recently, it was believed that only aphid-borne potyviruses affecting sugarcane and other groups of Poaceae were the mosaic and dwarf-mosaic viruses. Nevertheless, new criteria for the categorization of these groups of Potyviridae have been identified. They consist of the amino-acid chain makeup of their covering proteins, and serological connections with antisera resulting from epitopes within the amino-terminal, instead of the coat protein's central region.

Figure 2.11: Sugarcane Mosaic Virus mediated infection.

Source: http://blogs.ifas.ufl.edu/miamidadeco/2018/01/30/sugarcane-mosaic-virus/

Global distribution:The Sugarcane Mosaic Virus has been discovered in major sugarcane-producing states around the world, with the exception of Mauritius, Madeira, Mozambique and Guyana. There has also been a recent discovery that the sorghum mosaic virus, may also trigger symptoms similar to the mosaic of sugarcane. This particular infection is more common in India and Japan, where it was previously referred to as strain H of the SCMV virus, even though the assumption has already been disapproved by science.

Signs of disease:This disease causes systemic contagion in sugarcane plants, where the entire plant, including roots, might have traces of the virus. Nevertheless, the mosaic and necrosis symptoms are specifically seen on the leaves, though sometimes on stems as well. In some rare instances, the whole plant could become stunted.Otherwise, some of the classical signs of SCMV virus infection are diverse blends of green on the leaf surface, including paler green to beige chlorotic areas. The specific symptoms seen would depend on factors like the virus strain, environmental conditions, temperature and host cultivar (Emerson, 2019).

2.6.4 Maize Mosaic Virus

The maize mosaic virus belongs to the genus Nucleorhabdovirus and family Rhabdoviridae. It was first recorded in Venezuela during the 1960s and still remains commonplace in tropical and sub-tropical parts of Africa, Hawaii and the Americas, where its main transmission vector is the maize plant-hopper. The disease is common where sequential over-lapping maize plants are grown. Some of its telltale symptoms include; yellowing of leaf veins, though not the leaf mosaic which surprisingly remains whole. Growth retardation and loss of yield are also quite common in this disease (Emerson, 2019).

Virus makeup:The MMV particle comprises of a unipartite, single-stranded negative sense ribonucleic genome, including a pack of bullet-shaped, membrane-tied virions. MMV is usually transferred in a constant propagative way by P. maidis, plus the virus is not seed conveyed to the target cell. Other common hosts of the MMV virus include; itchgrass, teosinte, Zea and plains bristle grass, though maize and teosinte continue to be the main reservoir hosts.

Transmission method:PSV is transferred from one plant to another through various types of aphids, such as A. spiraecola, Aphis craccivora and andMyzuspersicae. It may further be transmitted through mechanical inoculation. In peanut seeds, it's also been found to be spread through seeds at very low levels, though this is not regarded to be very significant in virus spreading. The organism can further be hosted into a vulnerable field crop through aphids coming from nearby reservoirs, where other perennial host plants like alfalfa and clover are found. Then further spread into the farmyard by aphids. Also, it can be dispersed in perennial plants during harvesting, or possibly through root grafts as well.

2.6.5 Peanut Stunt Virus

Peanut Stunt Virus, also known as PSV, is a plant-based pathogenic virus belonging to the Bromoviridae family. It's also part of the Cucumovirus genus, which covers the cucumber mosaic virus as well. The molecule has a circular or isometric shape. This was first discovered in 1966, where it was known as Arachis hypogaea and considered to be a new disorder for peanuts and groundnuts. Some of the signs that infected plants showcase is severe stunting or dwarfing, thus the term peanut-stunt virus.

Symptoms of this disorder vary based on the host crop, including the virus strain in question. For groundnuts, there could be varying levels of stunting, petiole shortening, minimized leaf size, pod malformation and slight mottling. In beans, chlorotic mottling or mosaic happens mostly on trifoliate leaves, though in some cultivars these trifoliate leaves may appear lengthened and malformed. As for clovers, the plant may develop faint or severe mosaic, which could sometimes even include necrosis and leaf malformation. Whereas tobacco affected by peanut stunt virus may develop chlorotic spots, severe mottling, oak-leaf line marks and general chlorosis.

Prevention and management: Since PSVsarenot an economically vital condition in groundnuts, currently there aren't many control measures available. Nevertheless, it's still recommended to use cultural operations like removing the basis of infection and applying great quality seed to avoid disease spreading. As for forage legumes, like clover which arefound in perennial pastures, utilization of resilient or tolerant species is highly recommended.Within the community, PSV has been shown to be a weakening agent, especially in white-clover plants, since it renders them prone to damage or fatality from ecosystem-based stress and diseases.

2.7 PREVENTION AND CONTROL OF SOIL-BORNE VIRUSES

2.7.1 Control of Insect Vectors

Vector control refers to any method used to limit or eliminate the insects responsible for transmitting disease-carrying soil pathogens. Plant-based infections are usually spread by a wide array of organisms like insects, fungi, mites, plasmodiophorids and nematodes. Usually, the infection cycle starts with the vector meeting the virus within the soil or plant, then the virus is assimilated by the vector. It should then persist inside or on the insect for long enough until the virus is transported onto a new host, or delivered inside the plant cell.

Plant viruses depend on vectors for infiltrating through the host-cell membrane. Specifically, it's the viral capsid and membrane glycoproteins that are needed for transmission and factors of vector's specificity. Nevertheless, there are scientific ways you can manage insect vector transmission, without causing any harm to your plants or the soil itself. These include the following:

Blocking virus diffusion using glycoproteins and viral capsid proteins: As highlighted above, viral proteins are needed for attachment and penetration into the bug vector. Hence, exploiting these molecules for their particular binding properties to vector tissues could be a great approach for obstructing virus acquisition and relay. Genomes of crop viruses are rather small, and describing the viral attachment compound has been achieved using diverse and apparently intractable virus–vector methods. Using this information, recombinant techniques can be applied to minimize the spread of viruses by obstructing virus binding, including subsequent diffusion within the vector, and decreasing the vector population by using the virus protein to convey toxic substances to the insect (Emerson, 2019).

RNAi method: RNAi is a groundbreaking insect control technique that may also be applied todirectly target insect vectors. It's regarded as the foundation for the future blend of transgenic plants useful for insect pest management. As a virus prevention technique, RNAi is a pretty impressive option for plant-feeding bugs, since dsRNA is transferrable through transgenic expression in crops, transiently-expressed by virus vectors, or exogenously-spread by foliar sprays or soil drench to crops.For insect management, silencing of the bug gene through endogenous RNAi molecular machinery is activated by provision of dsRNA carrying the same sequence found in the genome transcript into the organism's cell. Upon entering, dsRNA is identified by the system as an RNAse III catalyst that attach the RNA into brief interfering RNAs. In insects, complete silencing occurs once the dsRNA chemical signal diffuses into the insect's body, through cellular uptake processes which are still undergoing investigation for different insect pest types.

Additionally, specific molecules found in vectors interrelate with the virus, plus while there are just a few-identified and non-identified receptors, the candidate recognition particles are further getting explored in these mechanisms. Because of the specificity of germ transmission by vectors, certain defined steps exist that symbolize great targets for disrupting the infection cycle. Integrated pest management is also a useful way of managing soil viruses spread through insects. It adopts strategies like plant-resistant cultivars, or plant seed and other propagative substances that are virus-free. You may also try removing and controlling reservoir hosts like over-wintering weeds, or avoid growing susceptible crops in places with a history of recurrent virus problems (Emerson, 2019).

2.7.2 Removal of Alternate Hosts

A considerable amount of crop viruses in multiple genres are shielded in and transferred through the soil. These microorganisms can be found all over the globe and many of them cause considerable economic losses to farms. Apart from known insects, other alternate hosts are fungal-resembling organisms and plant-parasitic nematodes. Still, some viruses may also be projected from diseased roots and transferred abiotically, without the need fora vector aid.

Disruption of alternate plant pathogen hosts: Scientists have recently discovered other viral hosts like arthropod-borne plant pathogens, which can also carry disease. For this particular case, recent focus has been on obstructing the spread of arthropod-carried pathogens through scientific methods. For instance, the pathogenic microorganism, Xylella fastidiosa, is usually transferred by leafhopper vectors, plus retained inside the vector foregut. Different from non-circulative crop viruses, X. fastidiosa organisms are usually attached to the foregut and subdivide within the insect. This processis commonly referred to as non-circulative propagative diffusion. Similar to plant viruses, bacterial cells which attach to the fore-gut are usually lost in the insect molting process (Emerson, 2018).

Studies have been carried out towards determining the bacterial factors of the interaction, identifying afimbrial adhesins as a key player in pathogen fixing to the vector. Corresponding research using antibodies on various microbial cell-associated proteins and particles confirmed the function of afimbrial adhesins in diffusion.Also, competition experiments using carbohydrate-binding lectins affected the interaction apparently by attaching the carbohydrates found inthe foregut attachment point and obstructing X. fastidiosa attachment. This test supports the theory that pathogen preservation in bug vectors is facilitated by specific connections and highlights similarities in vector transportation of different forms of pathogens.

Types of alternative hosts:

- Chytrid fungi: Commonly known as chytridiomycota, these are the only main group of original chitin-walled fungi thatproduce flagellate zoospores. Every other land-based fungushavelost its flagellate stages throughout the life-cycles, but chytrids have maintained them, requiring water for their generation and dispersal. Most plant viruses transferred through such fungi are carried by Olpidiumspecies; these are obligate crop parasites present mostly on the roots of both agricultural and wild plants all over the world. Moreover, Olpidiumbornovanus has lately been found to trigger root-disease independently, though most species are actually symptomless parasites that don't cause any damage by themselves. Instead, they are vital vectors of soil-based plant viruses found in the pathogen families of Tombusviridae and Ophioviridae.

- Plasmodiophorids: Plasmodiophorid virus vectors belong to the genera of Polymyxa, which mostly consists of P. graminis and P. betae. Nevertheless, the Plasmodiophorid classification is quite expansive and updated constantly as new particle phylogenetic information becomes available. Whereas they were once considered fungi, currently they're classified among Rhizaria organisms, which are found within the Cercozoa group. All Plasmodophorids have resting spores thatcan survive in the ground for several years, and these grow to generate water-based motile zoospores.They get attached to the host's root-cells and infiltrate the cell wall, delivering their cytoplasmic components into the host thus transferring any viruses they convey. These microorganisms are intracellular, root-infecting components that mostly affect cereal crops.

2.7.3 Sanitation Techniques

Proper sanitation is an effective tactic for effectual disease management in agriculture. It involves strategies or practices that seek to prevent the spread of viruses, by removing infected or asymptomatic contaminated tissue, including decontamination of farming tools, equipment and cleaning of hands.

Sanitation in the farm should occur before, during and right after the planting season. Before planting, adjust your weed management strategy so as to manage contamination. Most weeds are primary reservoirs for crop pathogens and bugs. By managing weed populations in the fields, it's possible to minimize pathogens and their infectious populations.Additionally, you should always wash soil residue from farm equipment before use, including scrubbing off soil components from shoes. These steps are particularly important for preventing the translocation of soil-borne pathogens, like Sclerotinia sclerotiorum which is a known agent of White fungus. This is inclusive ofothers such as Verticillium dalhiae, Phytophthora capsici, and various strains of Fusarium.

Generally, a power washer may be vital in fighting against plant-based diseases. Tools like discs, cultivators and plows must be power washed between sessions to remove soil that could possiblycarrying viral particles. Even more, avoid working fields whilst the plants are still wet. This procedure reduces microbial contamination from unhealthy plants to wholesome ones. For instance, this approach is crucial for tomato bacterial infections like spot, canker and speck which can prove difficult to control within the field.

Another strategy is removing diseased plants or specific parts. Immediately if symptoms emerge, gather, bag and extinguish or pile unhealthy material far from the fields. Eliminating infected fruit and crop materials from the farm can diminish the ratio of pathogen inoculum, which could have otherwise moved into unspoiled plant parts.

Figure 2.12: Removal of plant waste.

Source: https://4waste.com.au/rubbish-removal/remove-green-waste-home/

Ideally, discarded piles must be put away from growing fields and watercourses and, where possible, protected with an artificial tarp to fast-track microbial breakdown and curtail pathogen particles from escaping. Set aflame, chop and scatter, or deep plant the remnants at the season's end. The ultimate choice of elimination depends on the particular virus that was existing in field during the preceding season (Emerson, 2018).

During harvest, take care to only pick the healthy yield and avoid unnecessary mechanical destruction on the produce or other plant sections harvested. Injury, such as tiny wounds or cuts, maybe the entry point for germs that may affect fruit quality during transportation or storage.It's also recommended to inspect all crops before planting on the farm. Sometimes, plant material could carry infections and bug pests, meaning that introducing them to new fields may affect other crops that are still healthy.

Therefore, instruct field workers on the best ways to identify common disease signs and vermin that act as vectors. Checking often and carefully is necessary for identifying problems immediately they emerge. The more scrutiny is available for looking at your crops in the garden or field, the greater the chances that problems can be identified before they get out of proportion. Similarly, remember to disinfect knives, clippers and other farming tools regularly. To achieve this, cleanse tools with some detergent water, then dip or wipe in ethanol for disinfection purposes is must needed (Emerson, 2018).

2.7.4 Use of Certified Virus-free Seed or Stock

Choosing and planting clean, infection-free seeds is a key step in ensuring successful yield, no matter your acreage's size. Farmer also spend much less on seeking treatment for infected fields since the majority of plants shall grow to become healthy.

The significance of planting virus-free seeds: Virus diseases are typically transmitted from one planting season to the other through recycled seeds and tubers. Hence, researchers recommend the planting of illness-free seed as a crucial step in disrupting the infection cycle and stopping the syndrome from getting established. Even though it could be tempting to hoard a few seeds from your previous harvest each year, or buy seeds from an unapproved source, obtaining expertly grown, clean seeds shall pay long-term results. For all farmers, small and large-scale, a single infected seed usually has the potential to spread illnesses from one plant-to-another, including across different fields and farms. Most states have active seed approval programs, which allow producers to have their seeds examined by third parties to guarantee buyers that their produce would be of top quality. The seed inspection process usually begins when the seed is still being matured. Field examination may include checking the leaves and stem for illness symptoms, and guaranteeing that the seed yield is sufficiently isolated from illness sources. Right after harvest, it would undergo an inspection, testing and lab tests for additional evidence of illness.

Benefits of virus-free certified seeds: Since these seeds undergo thorough research and breeding processes, where plants are typically grown in controlled greenhouse environments, plus isolated from other crops. The chances of any virus passing through undetected arequite minimal. Furthermore, for open-pollinated crops like potatoes, there's lesser probability that seeds coming from such plants will mature true to type. Crops raised from such second-generation seeds usually bare smaller produce, which may also be affected by viruses transmitted from the parent plant.

However, certified seeds have been carefully selected through research to remove disease, as well as produce healthier produce that the farmer would be proud of. Starting your planting season with illness-free seed stock is usually the first step towards establishing a great stand thatshall enhance yield potential. There are numerous virus diseases which farmers should be aware ofwhen choosing the ideal seeds which will generate dynamic uniform stands, particularly the common mosaic is also known as virus Y, as well as late blight disease that mostly affects tubers (Gareth et al, 2018).

In particular, virus Y is very common in tubers and shows symptoms like wilting, mild foliar mosaic conditions or in worse case scenarios the crop may wither away. Likewise, certain strains of virus Y, like the necrotic variety, have been known to trigger necrotic symptoms just like the name suggests. Similarly, aphids are known vectors of the leaf-roll virus and among the primary means by which infections spread from one plant to another. Planting resistant seed varieties is among the most effective ways of dealing with aphid caused diseases. These crops can help minimize the rate of infection and virus transfer while maturing and yielding efficiently despite the contamination threat from a pathogen.

2.7.5 Growing Resistant Crop Varieties

It's estimated that viruses can reduce crop yields by up to 3 percent, whereas fungi and bacteria have a much higher loss rate of 15 percent. For some plants, like potatoes, the destruction caused by bacterial contagion is believed to be around 30%. To provide a remedy, researchers are considering the use of chemical agents that make plants infection resistant, including altering the genetic configuration of plants to make them resilient to wide varieties of microbial infections.

There are various ways of obtaining virus-resistant crops. These may include introducing the strain from an external source, careful selection, and impelled variation. All these methods may be applied at different phases in a continual process; for instance, varieties that are free from harmful bugs or plant symptoms may be availed for comparison with native varieties. More promising strains or lines may then be chosen for further transmission, additionally they may further be improved by encouraging as much disparity as possible via hybridization or specialized treatment. Lastly, the selection of the crops showing greatest signs of resistance takes place (Gareth et al, 2018).

Genetic engineering and breeding:Genetic manipulation of disease-resistant crops is usually a continuous process. Curated treatments for promoting gene changes may involve the application of mutation-sparking chemicals and radiation, using X-rays and ultraviolet (UV) treatment. These treatments typically prompt deleterious genomic changes, but, sometimes, useful results could also occur.Generally, the techniques used for breeding crops for virus resistance are alike to those employed in breeding for various other characteristics, only that it's a pair of organisms involved, which are the pathogen and host plant. Therefore, it's important to find out as much detail as necessary concerning the nature of genome inheritance of the disease-proof traits present in the host crop. This is including the existence of biological races or breeds of the pathogen.

Figure 2.13: Genetic engineering to control the mosaic virus in tomato.

Source: https://de.slideshare.net/SECBIO/genetic-engineering-13933607

Conventional breeding techniques play a crucial role in plant improvement, though they typically involve growing and assessing large crop populations over several generations, which could be an extensive and labor-intensive procedure. Alternatively, genetic engineering refers to the direct adjustment of a plant's genetic material through biotechnology, in order to make it more resistant to infections. Genetic modification has various advantages over conventional breeding. Initially, it allows for the introduction, eradication, modification, and adjustment of particular genomes of interest, but with reduced undesirable alterations to the remaining crop genome. Therefore, crops showing desired agronomic characteristics may be obtained in lesser generations, in comparison to conventional breeding. Additionally, genetic engineering permits for exchange of genetic matter across species (Gareth et al, 2018).

Accordingly, the raw genetic composition that can be worked on for this procedure is not only restricted to available genes found in the species. Plus, crop transformation during genetic processing allows for the introduction of fresh genes into tuber propagated crops like banana, potato and cassava. All these factors make genetic modification an essential tool for improving resistance against crop pathogens. Mostly, plant genetic engineering relies on predictive transgenic methods or genome-editing technologies for achieving the necessary results. For transgenic methods, genomes that encode preferred agronomic traits are implanted into the gene at random sites via plant transformation. These strategies usually result in strains having foreign DNA that are more resistant (Gareth et al, 2018).

2.7.6 Plant Quarantines

Plant quarantine can be described as the lawful enforcement of measures meant to inhibit viruses from spreading, or preventing further proliferation in case they have already entered a restricted zone. There are various methods of infection control that can be applied including; exclusion, elimination, resistance, protection, therapy and physiological control.

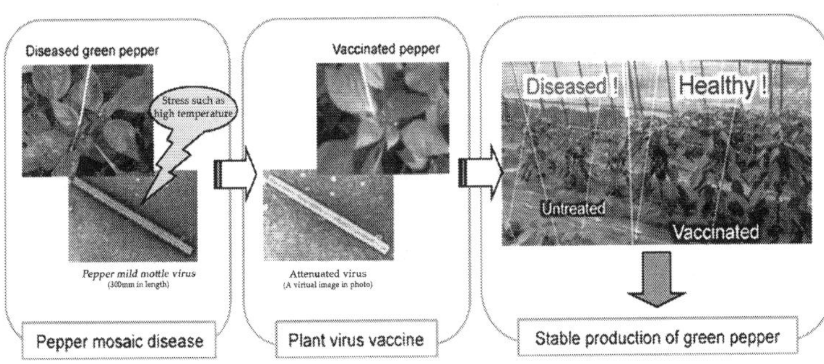

Figure 2.14: Plant quarantines to keep viral disease at bay.

Source: http://plantpath.ukzn.ac.za/Homepage.aspx

Exclusion or 'keeping off' is crucial to the notion of plant quarantine, whereas eradication techniques are applied to remove a newly identified pathogen. As such, plant quarantine could also be termed as, rules and policies promulgated by authorities to manage the introduction of crops, planting supplies, plant products, living organisms and soils. With the intention of preventing the involuntary introduction of foreign pests, weeds and microbial substances dangerous to the farming or the ecosystem of a region, and when introduced, to avert their foundation and further spreading.

Plant quarantine, as such, is developed as a protection against destructive pests or pathogens which are foreign to a region (Gareth et al, 2018).

The significance of placing restrictions on the movement of virus-infected plants was first discovered in 1860, when grapevine phylloxera was brought to France from the U.S. The plant caused tremendous devastation to local crop population, leading the government to place restrictions on its use and plantation.

Virus detection method:During quarantine, serological tests are usually done to detect and identify viruses which may be a threat to plant populations. The phage-plague procedure is quite effective in identifying pathogens and their respective strains. Indicator test crops may as well be helpful since they might show pathogenic races or strains within certain species of viruses.

Alterations of the universal incubation tests have further been applied for the detection of particular plant pathogens. Similarly, deep-freezing blotter exams and 2, 4-D blotter assessments have been crucial in the discovery of black-leg microbes in crucifer crops.

As for vegetative propagules, laboratory techniques may be sufficient for detecting insects, mites, bacteria and fungi. Nevertheless, for the identification of systemic fungi pathogens, viruses and viroids, isolation planting for a season, year or longer in quarantined net or glasshouses is necessary. The availability of these protective structures in massive numbers may be a costly proposition, though the quarantine protections provided by them towards any nation are worth that expense.

Quarantine methods:

- Heat treatment: Heat treatment, also known as hot-air treatment or boiled water treatment, is commonly applied in the quarantine viruses, fungi and bacteria. The basic principle involved is that treating temperatures should be considerably high to eliminate the associated microbe, though not the host plant. Nevertheless, in most instances, the safety margin is quite narrow and, hence, the temperature must be very precisely controlled. For viruses, the recommended hot-water treatments administered should be between 45-50°C.
- Fumigation: Fumigation of infected material under standard environmental or reduced pressure levels is effective as a quarantine remedy against insect carrying virus vectors. For this

purpose, fumigants such as HCN, phosphine, methyl bromide and Ethylene dichloride-carbon tetrachloride (EDCT) are commonly applied.

- Chemical treatments: Chemicals fumigants may be spread as dust, spray, dip or slurry. It must be guaranteed that the dosage of chemical used is only enough to eliminate the microorganism, but without damaging the host just like in the other quarantine processes discussed above.

Chapter 3 Soil Bacteria

CONTENTS

3.1 INTRODUCTION

Soil bacteria are single-celled organisms living in the soil. So very tiny, soil bacteria are found in millions. For instance, a teaspoon of soil contains around 1 billion bacteria. Soil bacteria just like soil fungi may be classified into their major functional groups.

- The first group involves decomposer soil bacteria. They consume plant residues and their exudates from roots or even plants litters. They also participate in the breakdown of pesticides and other pollutants in the soil.
- The second group of soil bacteria involves mutualistic soil bacteria. Their main role is to establish a mutual relationship with plants. They relate mutually with plants in symbiotic or non-symbiotic relationship. A good example is nitrogen-fixing bacteria.
- The third group of soil bacteria involves pathogenic soil bacteria. They act as parasites on plants and may cause diseases and other devastating effects on plants if not checked.
- The last group consists of lithotrophic bacteria. They mainly obtain their energy from iron nitrates and other compounds instead of carbohydrates from plants. Their main role is breaking down pollutants and nitrogen cycles.

It is important to know the soil bacteria from various functional groups are involved in water retention, movement of nutrients and disease suppression. For instance, a soil bacterium that produces substances that bind the soil particles will consequently improve soil stability and water infiltration and retention in the soil.

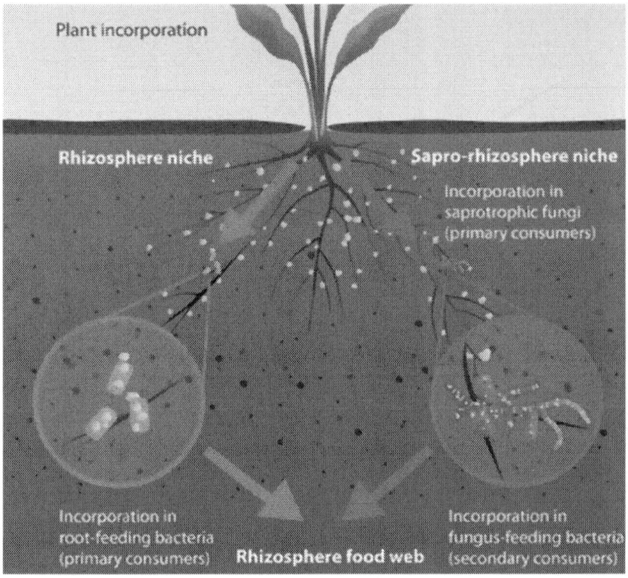

Figure 3.1: Presence of bacteria in soil environment.

Source: https://content.iospress.com/articles/journal-of-intelligent-and-fuzzy-systems/ifs179083

3.2 CHARACTERISTICS OF SOIL BACTERIA

Soil bacteria are almost similar in size with small clay particles while others are big up to the size of silt soil particles. This explains why soil bacteria occur in various sizes and shapes. They therefore grow and live as water films within soil particles and root' srhizosphere. The small sizes allow soil bacteria to live and adapt rapidly to changing ecosystems much faster than other soil microorganisms that are competing with bacteria for space. However, most soil contains those bacteria cells due to their high population. Soil bacteria will reproduce when soil, water and environmental conditions become more favorable but their numbers double rapidly in no time into millions in less than 30 minutes. This explains why bacterial populations in some areas known as hotspots are found in billions while in some areas, they are found less than 1 million (Hoorman, 2011).

Soil bacteria have no nucleus. They contain a continuous double strand of DNA. They have an irregularly shaped nucleoid without a membrane. Moreover, they have a cell wall thatprotects them.Most soil bacteria reproduce through binary fission where cells are split after reaching a certain size. This reproduction is mainly asexual where two daughter cells fuse. This leads to transmission of DNA from foreign soil bacteria to other daughter cells.

Soil bacteria are found in various shapes. Some are spherical and known as cocci. Spiral bacteria are spiral in shape. Other soil bacteria maybe rod-shaped and are known as bacilli bacteria. Others may be short or long filaments while some may be star-shaped or squared in shape.Bacteria in the soil may also be characterized based on their feeding styles, for instance, heterotrophic soil bacteria feed on organic carbon from organic matter. They are commonly found in compost. An autotrophic soil bacterium makes itsown food through photosynthesis by using sunlight water and carbon dioxide. Additionally, soil bacteria may further be characterized into being aerobic or anaerobic based on how they undertake metabolism. For instance, anaerobic soil bacteria are found in ecosystems without oxygen supply but their counterparts known as aerobic bacteria need oxygen to survive and are commonly found in places with sufficient supply of oxygen (Hoorman, 2011).It is therefore interesting to note that some soil bacteria may be facultative anaerobes or Aerobes. This is a special feature of certain soil bacteria that are able to survive soils with or without oxygen supply.

Finally, it is useful to characterize bacteria based on the structures of their cells. Soil bacteria cells are characterized by the following main features. Soil bacteria cells have; a capsule layer, a cell wall, a plasma membrane, cytoplasm and DNA. All these special characteristics explain why bacteria cells are different. They are normally prokaryotes.

Soil bacteria live in different microenvironments and food sources. For instance, they are more prevalent with increased availability of substrates such as plant residues and substrates found around the root surface in the soil.Soil organisms particularly soil bacteria determine the growth of plant communities thatgrow in a particular ecosystem. In some plant ecosystems, soil bacteria may encourage specific types of plants to grow while at the same time discouraging other plant species from growing (Hoorman, 2011).

3.3 THE ROLE OF SOIL BACTERIA IN THE SOIL

Soil bacteria perform a number of functions in the soil which are both useful to the health and growth of plants soil stabilization and also the health of other microorganisms in the soil.

- To begin with, soil bacteria, produce hyphae-like products which help in binding soil particles together. The bonded soil particles help in retaining water in the soil. Moreover, the same soil particles are more stable in structure.
- Secondly, soil bacteria assist in the decomposition of dead organic residues in the soil and even on the ground surfaces. The end products of this breakdown process result in production of mineral nutrients thatare useful to plants because they boost soil fertility. Consequently, the health of plants and that of the soil is improved therefore soil bacteria are important recyclers of soil nutrients from dead organic matter.
- Furthermore, soil bacteria, particularly rhizobium, establish mutualistic relationships with plants. For instance, bacteria hosted by soybeans form nitrogen nodules which are useful in nitrogen fixation for host plants. This type of bacteria converts atmospheric nitrogen into nitrogen compounds that can be easily taken up by plants.
- Denitrifying soil bacteria helps reduction of excess nitrogen compounds in soil by allowing nitrogen to escape back to the atmosphere. However, there is another species of soil bacterium which trap nitrogen from atmospheric nitrogen and convert them back to the soil when there is a limited supply nitrogen compounds in the soil. This type of bacteria is called nitrifying bacteria.
- Pathogenic soil bacteria are most dangerous in the role of causing diseases in plants such as bacteria blights. However, in some cases, some soil bacteria may act as disease suppressors and assist the soil to be more suppressive to disease-causing microorganisms. This way they protect plants from dangerous microorganisms such as some soil fungi and other harmful bacteria from causing diseases.
- It is also important to know that bacteria play a very important role in the degradation of pesticides that would have otherwise contaminated soils and the environment. After breaking down pesticide particles less harmful products are produced which does not harm the soil or the environment.
- Soil bacteria also contribute to the population and the existence of other soil microorganisms. Bacteria and other soil microorganisms

compete for nutrients and food, this way bacteria regulate the numbers, reproduction and growth of other soil microorganisms.

- Furthermore, soil bacteria produce waste which is useful in cementing sand, silt and clay soils. This leads to soil stability in terms of structure and properties. This way soils will generate new plants faster mainly because bacteria would have provided a foundation for the establishment of new plants in the soil.
- Finally, soil bacteria play a role in altering conditions to allow the growth of specific plant species. Without bacteria in the soil most weeds struggle to exist in the newly cultivated farms where new soils are forming. It is thus important to allow bacteria exist mutually or either independently in soils.

3.4 TYPES OF SOIL BACTERIA

3.4.1 Nitrogen Fixers

Nitrogen-fixing soil bacteria is a group of soil bacteria capable of extracting nitrogen gas from the atmosphere then converts it into nitrogen compounds and nitrates that can be easily assimilated by the plants.Rhizobia soil bacteria, for instance, traps atmospheric nitrogen in air and fixes it to plants. They mainly live in special roots sections of legumes in plants such as beans and soybeans among other hosts. Once they enter the roots of the legumes, they will cause excess production of plant tissue around it. This results in formation of tubercles (Hoorman, 2011).

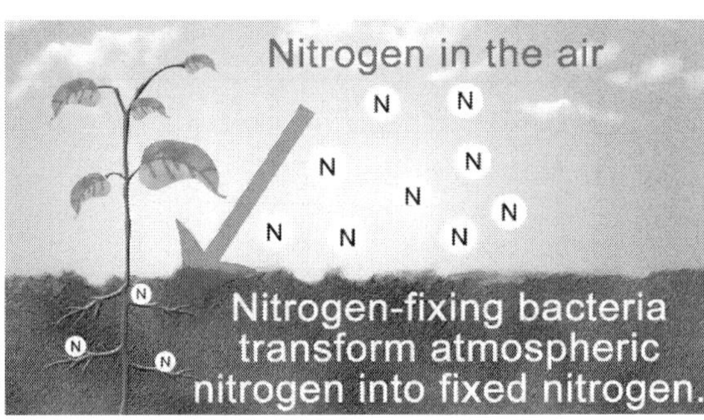

Figure 3.2: Nitrogen fixing bacteria in association with plants.

Source: https://biologywise.com/nitrogen-fixing-bacteria

Nitrogen fixing bacteria usually enter the roots of legumes via the rhizosphere in their attempt to access carbohydrates from plants. This usually happens when plans are weak to resist nitrogen fixing bacteria from entry. Traditionally they come to the plants as problem solvers to fix nitrogen deficiency in plants.

Agriculturalist introduces nitrogen fixing bacteria through inoculation of seeds. Germinating crops from these seeds will have constant nitrogen supply courtesy of inbuilt nitrogen fixers.Once they are fully settled in the plant's legumes nitrogen fixers will establish symbiotic relationships with host plants so as to excess carbohydrates while fixing the poor supply of nitrogen nutrients in plants. This is the reason why nitrogen fixers are at times referred as mutualistic bacteria. It is therefore important to note that various conditions affect the nitrogen fixation process by soil bacteria. These factors are closely related to those that maintain the growth and the health of plants. However, generally the amount of nitrogen generated by legumes depends on specific plant species, soil properties and farm management practices.

The presence of potassium in the soil plays a very critical role in nitrogen fixation. Potassium is useful for the growth of roots and promotes development of big nodules on roots. Therefore, adequate supply of potassium is an important ingredient proper and continuous nitrogen fixation by soil bacteria (Hoorman, 2011).

Generally, biological soil nitrogen fixation by bacteria results in the production of ammonia which can be readily observed by plants in various forms. Soil bacteria utilize an enzyme nitrogenase as catalyst so as to convert nitrogen in the atmosphere into ammonia.

Many soil bacteria fix huge amounts of nitrogen without any direct relation to other organisms. Free-living bacteria in anaerobic conditions take up the role of fixation while avoiding oxygen which may inhibit nitrogen fixation. This process, however, is known as fixation by free-living bacteria. On the other hand, associative nitrogen fixers, fixation occurs when bacteria especially rhizobia establish mutualistic relationships with plants. Rhizobia fix huge amount of nitrogen within the plant'srhizosphere. This condition is usually witnessed in many legumes which are planted alongside other crops so as to encourage nitrogen fixing bacteria (Hoorman, 2011).

In conclusion, it is important to note that though plants may be supplied by nitrates through fertilizer, the role of nitrogen fixers cannot be overlooked. They play a very crucial role and if not well checked their poor health and well-being will greatly impact plants. Agriculturalist should adopt practices that encourage the multiplication of nitrogen fixing bacteria i.e.by growing number of legumes.

3.4.2 Nitrifying Bacteria

These bacteria converts soil nitrogen into forms that can be accessed by plants. These nitrogen compounds include ammonia and nitrates. Nitrification is an important process that boosts soil properties and at the same time contributes to oxidation of methane and breakdown of organic compounds and carbon dioxide.

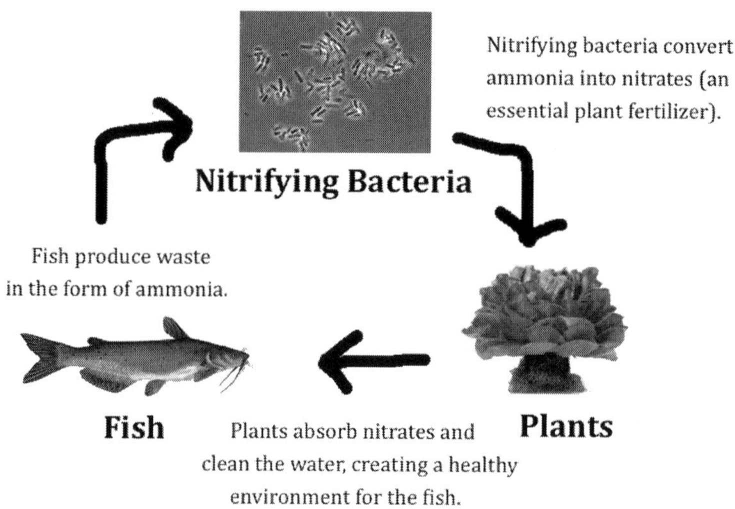

Figure 3.3: Nitrifying bacteria and their association with plants.

Source: https://www.pngfly.com/png-2p2xnw/

Nitrifying soil bacteria are mainly obligate aerobes that require oxygen for oxidation. Examples of these bacteria are Nitrosomonasspp. This type of bacteria gets energy from ammonia when converting it to nitrite. On the other hand,Nitrobacterspp. will process nitrite into nitrates (Aislabie&Deslippe, 2013).

When conditions are favorable such as moderate warm temperature and good water content; nitrification occurs in the soils. Nitrification also occurs if ammonia sources are readily available such as through fertilizer, compost or urea.Once nitrification happens plants roots will readily absorb the nitrogen compounds produced during nitrification. Nitrification is common in warm temperate soils but conditions below 5°C will discourage nitrification process. Furthermore, well-aerated soils favor nitrification mainly because nitrifying bacteria use oxygen to oxidize ammonia.

Nitrifying bacteria are very few in the environment but very many in places with huge sources of ammonia. sewage and places with huge protein decomposition are hotspots to nitrifying soil bacteria. Nitrifying bacteria are also found in aquatic environments that have high ammonia content. Nitrifying bacteria play a lot of roles in agriculture and therefore should be encouraged using various methods. For instance, soils may be injected with nitrifying bacteria so as to help boost the nitrification (Aislabie&Deslippe, 2013). Furthermore, farm practices such as providing soils with compounds that contain ammonia should be encouraged. This is because ammonia is an important source of nitrification. For this case manure and urea are added to the soil so as to provide the soil with a steady supply of ammonia which is very critical in the process of nitrification by soil nitrifying bacteria.

In places with the dangers of nitrification soil nitrifying bacteria may be discouraged and controlled mainly by using inhibitors of soil nitrifying bacteria. Inhibitors of soil nitrifying bacteria are combined with fertilizer formulations during manufacture. When this fertilizer is applied to the soil the inhibitors will act negatively consequently reducing nitrification in the soil. Furthermore, the water logging of farms will inhibit the growth of nitrifying soil bacteria due to poor aeration. As mentioned earlier nitrifying bacteria use oxygen to oxidize ammonia into nitrates does water logging will help reduce nitrification by cutting off the supply of oxygen, which is much needed by soil nitrifying bacteria in order to oxidize ammonia (Aislabie&Deslippe, 2013).

Furthermore, low temperatures discourage the reproduction and growth of nitrifying bacteria in the soil. Due to low temperature, nitrifying bacteria will be discouraged from reproduction and growth resulting in either their dead or migration and consequently reducing nitrification in the soil or water bodies.

In conclusion, it is important to note that nitrifying bacteria can be both useful and harmful to man depending on how man will like to utilize nitrifying bacteria in the soil or water bodies. In useful ecosystems nitrifying soil bacteria is an important soil bacterium that can be used to inject nitrogen compounds into plants.

3.4.3 Denitrifying Bacteria

Denitrification is the reverse of nitrification. Whereas nitrification by soil bacteria involves oxidation, denitrification mainly involves reduction. Denitrification by soil bacteria is a process where nitrates are converted to nitrogen gas and later sent to the atmosphere. Soil denitrifying bacteria use nitrates in their respiration instead of atmospheric oxygen. It occurs faster in warm soil and is more common in wet waterlogged soils where oxygen supply is limited.

When water levels in the soils exceed 60% denitrification by soil denitrifying bacteria will begin. The end products may be various gases such as nitrogen and nitrogen oxide. The lesser the supply of oxygen, the higher the rate of bacteria denitrifying more nitrates out of the soils (Aislabie&Deslippe, 2013).

Furthermore, denitrifying soil bacteria are mainly found in the topsoil. Their activity is at optimum here because many denitrifying soil bacteria live there. The survival in the topsoil is mainly because topsoil provides rich sources of carbon and other organic matter for the metabolism of denitrifying soil bacteria.

It is therefore important to know that the populations of denitrifying soil bacteria available in specific soil and specific areas are determined by several factors which include but not limited to the following;

- Availability of nitrates-Denitrifying bacteria occurs mainly in places where nitrate is readily available because they need nitrates to denitrify.
- Temperature-Denitrifying bacteria live in warm soil, therefore, any fluctuations in temperature will affect their growth, activity and reproduction in the soil.
- Soil moisture content-Wet soils will generally trigger denitrifying bacteria activity that's more than denitrification will occur compared to well-drained oxygenated soils.
- Availability of carbon compounds-The presence of carbon promotes the growth and the activity of denitrifies in the soil. Denitrifying soil bacteria need soluble organic carbon as their major source of energy for metabolism.

Though denitrifying soil bacteria may be useful for reducing excess nitrates in the soil if not checked may bring devastating effects to farm soils and the environment. They may lead to loss of nitrate fertilizer in the farm soils. Due to this fact, several farm practices aimed at reducing the populations of denitrification bacteria have been discovered. These practices include;

- Proper soil drainage so as to encourage proper oxygen supply to the soil thus discouraging the growth and activity of denitrifying soil bacteria.
- Discourage water logging in farms which will otherwise provide denitrifying bacteria with a favorable condition for denitrification.
- Use inhibitors to slow down the activity of soil denitrifying bacteria thus less exposure of nitrogen fertilizer to denitrification.
- Proper and timely irrigation with a less and constant supply of water helps in discouraging the activity of denitrifying soil bacteria.
- Avoid excess nitrogen fertilizer which may act as important sources of nitrates needed by denitrifying bacteria.
- Plant cover crops help reduce nitrates in between crops and consequently reduce denitrification.

In conclusion, it is important to note that several factors determine the growth, activity and the well-being of denitrifying soil bacteria in the soils. These factors are generally interrelated and all must be considered so as to understand denitrifying bacteria in the soil.

3.4.4 Sulfur Oxidizer

This is a type of soil bacteria which oxidize sulfur into sulfites which can be readily taken by plants. Most plants use sulfites for the production of many protein compounds.

Sulfur oxidizing soil bacteria are in various groups;

- Beggiatoa bacteria are a thread like soil bacteria which oxidizes hydrogen sulfide to get energy. Normally they store sulfur in their own cells and later when hydrogen sulfite isdepleted in the soils they will oxidize the stored sulfur. This bacterium does this without the aid of any form of light.
- The other group of sulfur oxidizers is the red-pigmented bacteria. This bacterium uses sunlight for growth. They are usually found in seawater near the shores.

Interestingly, several interrelated factors determine the growth and population of sulfur-oxidizing bacteria in the soil. To begin with sulfur-oxidizing bacteria exist in warm temperatures. This results in healthy growth and multiplication of these soil bacteria when presented with favorable temperatures for growth and reproduction (Hoorman, 2011).

Soil-moisture is also important in the activity and growth of sulfur oxidizers. Water is required for oxidation. Therefore, any chemical reaction that leads to oxidation of sulfur in the soil cannot start without the presence of enough soil moisture. Sulfur oxidizers are affected by the availability of sulfur compounds in the soil. When more compounds are present in the soil, the more likely sulfur-oxidizing bacteria will act on these compounds to generate sulfur nutrients for plants to assimilate. Most sulfur-oxidizing bacteria are found in soils with high levels of sulfur-containing compounds such as sulfur nitrates and salts. Other factors such as soil drainage and soil pH are also important factor for the survival of sulfur-oxidizing bacteria in the soil. For instance, when soil hasless oxygen supply thus most sulfur-oxidizing bacteria will not be found in this ecosystem but well-aerated soils will contain more sulfur-oxidizing bacteria because of the availability of oxygen to oxidize sulfur.

Finally, oxidizers depend a lot on the size of particles available in the soil. This is because large sulfur particles provide large surface areas for sulfur- oxidizing bacteria to act on them.

It is important to note that the amount of sulfur- oxidizing bacteria in the soil may be controlled through several farm management practices. These practices include the following:

- Practicing crop rotation to ensure availability of crops that can host sulfur- oxidizing bacteria. Moreover, their residues will provide important sources of sulfur nutrients.
- There is a need to use compost manure to provide sulfur oxidizes with a rich supply of nutrients.
- Proper and timely irrigation should be used so as to avoid water logging which may cut off oxygen supply which is much needed by sulfur- oxidizing bacteria in the soil.

In conclusion, sulfur oxidation by bacteria particularly soil bacteria have been applied widely in oil Fields so as to help reduce the number of sulfur contaminants in the fields.

3.4.5 Anaerobic and Aerobic Soil Bacteria

These two types of soil bacteria are arrived at based mainly on their need for oxygen. Soil bacteria that prefer well-aerated soils with constant supply of oxygen are referred to as aerobic bacteria. Anaerobic bacteria in the soil produce their energy without oxygen.

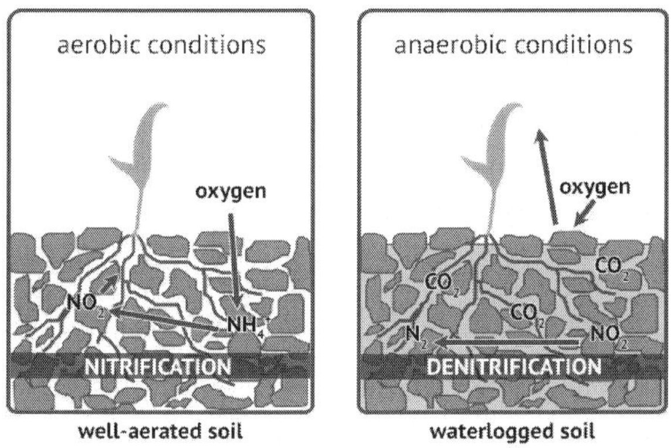

Figure 3.4: Anaerobic and aerobic soil bacteria.

Source: https://www.philpoteducation.com/mod/book/tool/print/index.php?id=914&chapterid=1333

Anaerobic soil bacteria: It is believed that most anaerobic soil bacteria produce their energy without oxygen. An anaerobic bacterium does this through Glycolysis without requiring any supply of oxygen for metabolism. However, some anaerobic soil bacteria may tolerate oxygen and actually use it sometimes.

Anaerobic soil bacteria may be classified based on their metabolic characteristics:

- Facultative anaerobes utilize oxygen for respiration in metabolism but insufficient oxygen will not hurt them
- Obligate anaerobe avoids oxygen and therefore they mainly undertake anaerobic respiration.
- Aero-tolerant anaerobes are not affected in any way by oxygen but oxygen makes them unable to carry out their metabolism.

In conclusion, anaerobic soil bacteria use pathways like anaerobic respiration or fermentation in energy metabolism. Moreover, anaerobic bacteria in soils are normally found in deep compacted soil particles that have limited oxygen supply. When in this depth, they push out aerobic bacteria out of the soil and that is why aerobic bacteria are normally found in manure mainly in the topsoil or humus.

Aerobic soil bacteria: This type of soil bacteria requires a constant supply of oxygen for their metabolism to occur. Aerobic soil bacteria are found in various soils and play roles such as nitrification and sulfur oxidation. For instance, convert ammonia into nitrates which can be taken up by plants. Additionally,oxygen is important for aerobic soil bacteria because oxidation of compounds in the soil requires oxygen to kick-start the chemical oxidation reaction process. Aerobes also use oxygen to metabolize and carry on fundamental processes.

Moreover, aerobic soil bacteria use oxygen to decompose most organic matter in the soil. Most processes of decomposition which involves the breakdown of dead organic matter are done by aerobic bacteria.Obligate aerobic bacteria require oxygen so as to start fermentation. This results in the production of hydrogen which enables aerobic bacteria use hydrogen as their basic source of energy. Furthermore, due to this high demand for oxygen aerobic soil bacteria are found in well-drained oxygenated soils thatare mainly found on the top layer where they play their final roles without any harm from aerobic bacteria or other soil microorganisms (Hoorman, 2011).

Most aerobic bacteria are widely used in the soil so as to enable nitrification and oxidation of sulfur. These processes are very important for the stability of the soil, plants and their environment. For instance, sulfur-oxidizing bacteria are needed to remove excess sulfur compounds in the soil which may otherwise contaminate the environment especially groundwater poisoning.

In conclusion, it is important to note that both aerobic and anaerobic bacteria have their harm and benefits in equal measure. It is therefore important to maintain anaerobic and aerobic bacteria in the soil because of their benefits both to man and to the environment outdo their harm.

3.4.6 Actinobacteria

This type of bacteria is commonly found in torrential and aquatic environments. The major role of these soil bacteria is to decompose organic matter into soil nutrients that can be absorbed by plants. Majorly, they act as organic matter decomposers.

Main features;

- Most actinobacteria are aerobic and unicellular who form aerial and substrate mycelium.
- Soil actinobacteria reproduce by binary fusion while producing spores.
- Soil can be described as being conical and compact in appearance.
- They can be classified into thermophilic, acidophilic and symbiotic actinobacteria among other groups based on their ecosystem's food and characteristics.
- Actinobacteria have a high content of cytosine in their DNA.
- They are largely a group of Gram-positive bacteria.
- Actinobacteria though they lack a cell wall they produce mycelia which areslender and has no septa.

Actinobacteria live in soils with very high sensitivity to low pH and acid. Studies show that most actinobacteria avoid environments that are believed to have low pH i.e. acidic environments. They also avoid desert ecosystems due to extreme conditions but they are very many in most arable torrential lands and forest soils.Most actinobacteria species found in water may have entered from adjacent lands leading to their living and adaptation to water and sediments in shores. Soil scientists and biologists argue that Actinobacteria are found worldwide in almost all ecosystems and in various types of soils but they vary in number (Aislabie&Deslippe, 2013).

Actinobacteria plays important roles in the soil which involves the decomposition of organic matter such as chitin and other compounds leading to replenishing of nutrients in the soil and humus formation. Actinobacteria decomposers target plant residues and organic compounds that cannot be broken down simply by other soil microorganisms.

Soil actinobacteria living in symbiotic association with plants get access to saccharine from plants. Actinobacteria do this by colonizing plant roots so as to fix nitrogen in other words they act as mutualisms. They supply plants with nutrients while they also have constant supply of carbohydrates generated by plants during photosynthesis.Actinobacteria such as Streptomycesspp are useful in the field of medical engineering. They are used to produce antibiotics such as amino glycosides and macrolides.

Moreover, actinobacteria from soils found in water bodies play a very important role in supplying aquatic plants with nutrients that are useful in their well-being. Some biotechnologists have engineered soil actinobacteria in the production of pesticides such as insecticides and herbicides. However, some actinobacteria are also used to reverse contamination caused by pesticides through biological degradation (Aislabie&Deslippe, 2013).

Finally, medical engineers have used the active metabolism of actinobacteria to produce antibacterial and drugs that are used to control and treat human and animal diseases.

Interestingly, Actinobacteria was confused with soil fungi because of its ability to produce mycelia and spores which are common characteristics of soil fungi. Therefore, further studies on this complex organism on the production of mycelia and spores need to be studied and analyzed in detail so as to clearly differentiate it from soil fungi.

In conclusion, it is important to note that these pieces of soil bacteria are unique compared to the other bacteria types such as nitrogen fixers. In this case, actinobacteria are used for several beneficial purposes.

3.4.7 Disease Suppressive Bacteria

This type of soil bacteria is used to suppress disease-causing pathogens in the soils and other ecosystems. For instance, Bacillus megaterium and Bacillus subtilis are used to suppress diseases in particular crops.

The protection of plants from pathogens by soil bacteria is largely aided by disease oppressiveness of the specific soils. Naturally, disease suppressive soils are believed to offer protection to plants but recent studies show that such soils contain disease suppressive bacteria that are actually involved in protecting plants against any form of infections (Doetsch& Cook, 2012).

It is for these reasons that; it is believed suppressive soil contains beneficial rhizobacteria known mainly as plant growth-promoting rhizobacteria (PGPR). This type of bacteria colonizes the rhizosphere of the roots and once fully established, these soil bacteria will boost the defense mechanism of the host plants.

Rhizobacteria protect plants from infection by pushing out or repelling parasitic disease-causing pathogens such as fungi and harmful bacteria within the plant tissues. It is important to note that this rhizobacteria process works closely with soil and plants so as to suppress pathogens either directly or indirectly.This is process produces suppressive metabolites which are used to provide defense mechanism against disease-causing pathogens in the soil. They also assist plans by providing nitrogen fixation and replenishing nutrients in the soil. These disease suppressors are normally mutualistic and they benefit from plants by accessing a constant supply of carbohydrates for their energy in metabolism.

Other disease suppressive bacteria in the soil such as actinomyces produce a lot of hormones to increase the uptake of mineral compounds by the roots. Most plants attract a lot of suppressive bacteria in their rhizosphere for disease control and to boost nutrient uptake (Doetsch& Cook, 2012).

It is interesting to note that disease suppressive soil bacteria perform other roles in the soil such as nitrogen fixation while others are involved in the decomposition of organic matter. Similarly, some decomposers act as disease suppressors by providing plants with important nutrients which if otherwise insufficient will lead to plants being susceptible to diseases. It is because of this that most soil bacteria are believed to have active and inactive roles.

Denitrifying soil bacteria contribute to disease suppression. By nitrifying excess nitrogen nitrates to produce atmospheric nitrogen they help plants reduce excess nitrogen compounds in the soil which may have caused infections (Doetsch& Cook, 2012).

Finally, it is important to adopt practices that boost disease suppressive bacteria. Disease suppressive bacteria in the soil may be encouraged through farm practices such as growing crops that are able to host these suppressive bacteria. For instance, growing legumes will greatly boost the population of rhizobia bacteria which makes plants more resilient due to the huge population of disease suppressive rhizobia.

In conclusion, it is important to note that disease suppressive soil bacteria are affected by several environmental and soil properties factors which if not checked will affect their population in the soil. Furthermore, continuous use of fungicides may affect the growth, reproduction and even the lifestyle of the suppressive bacteria in the soil.

3.5 BENEFITS OF SOIL BACTERIA

Soil bacteria are useful to man, plants, soils and our general environment in various ways.

- First, soil bacteria particularly decomposers are used by biotechnologists mainly as a bio-remedy to control soil and environmental pollution. Environmentalists and soil scientists use decomposer soil bacteria to initiate the biological breakdown of pesticides and other soil pollutants. This results in the production of less dangerous products that do not contaminate the environment or soil and even groundwater.

- Secondly, soil bacteria are currently used in pest control. Pathogenic soil bacteria, when given suitable conditions, will multiply and suppress other soil microorganisms which would have caused devastating effects on plants especially farm crops. In this case disease suppressing soil bacteria are introduced into the soil to help improve soil resistance to disease-causing pathogens.
- Air circulation is also regulated in one way or another by soil bacteria. Soil bacteria recycle their compounds such as oxygen and nitrogen thereby presenting living things such as human beings with suitable air circulation. An example is nitrogen fixing bacteria and the denitrifying bacteria. While nitrogen fixers trap nitrogen for plants, the later denitrify nitrates through oxidation into nitrogen gas which escapes back to the atmosphere.
- Moreover, soil bacteria are used in fiber retting. In this case, plants are immersed in water so as to swell and thereafter they are inoculated with soil bacteria which will hydrolyze plant cell walls and consequently separating fibers. In a more traditional way plants are left overnight on the ground so that soil bacteria will attack them when they are moist with dew leading to retting.
- In agriculture, soil bacteria are utilized in various aspects. Nitrogen fixers are used to fix nitrogen compounds into crops while decomposers are used to decompose plant residues and to improve soil fertility. Decomposers are also used to bond soil particles so as to improve soil stability and structure.
- Furthermore, some anaerobic soil bacteria are used in fermentation. When combined with yeast in anaerobic conditions, soil bacteria provide proper fermentation. This is generally applied in the fermentation of many alcoholic drinks and in the making of many organic compounds such as vinegar.
- Currently, biotechnologies use soil bacteria in the production of several products such as enzymes, perfume, ethanol, organic acids among others.
- Furthermore, some soil bacteria species are used by biotechnologists in the synthetic manufacture of amoxicillin and other antibiotics which are useful in controlling and preventing

human and animal diseases.

- Similarly, sulfur-oxidizing soil bacteria have been used in the oxidization of sulfur compounds in the soil so as to make it available for plants. For instance, sulfur-oxidizing soil bacteria oxidize sulfur into sulfite nutrients that can be readily absorbed by plants. This oxidation process has helped in reducing sulfur compounds which may contaminate soil and groundwater.
- Finally, genetic engineers use soil bacteria in the replication of DNA sequences in genes leading to transfer of DNA genes from parents to new organisms. This process has commonly been used in genetic engineering to produce insulin for diabetics.

3.6 MANAGEMENT OF SOIL BACTERIA

Soil bacteria and its population in the soil may be encouraged or discouraged depending on their uses or harm both to the soil and man. Several management practices have been discovered by man in an attempt to control and regulate bacteria in the soil.

3.6.1 Promoting useful Soil Bacteria

- Useful soil bacteria may be encouraged by addressing soil properties to ensure that soil bacteria have suitable conditions for growth and reproduction in the soil. Soil pH is useful in determining the growth of bacteria in the soil. Soil acidic levels should be checked so as to give bacteria a healthy environment for growth and reproduction.
- Secondly, soil bacteria can be encouraged by reducing tillage practices which may affect their growth, reproduction and lifestyles in the soil. For proper growth of microorganisms such as soil bacteria tillage practices should be improved so as to avoid disturbing the bacterial populations; their life cycles and their reproduction in the soil.
- Moreover, soil bacteria may be encouraged by providing enough substrates. Soil bacteria such as decomposers need substrates as their main source of food. This can be achieved by providing plenty of organic matter which when broken down, soil bacteria will have enough substrates for metabolism and at the same time, they will provide plants with important nutrients.

- Furthermore, mulching promotes soil bacteria. Mulching warm temperature and moisture around the crops will be encouraged. These conditions are suitable for the growth of soil bacteria. Additionally, mulching material will act as an important source of organic matter to soil bacteria.
- Finally, growing plants that encourage bacteria is useful in encouraging soil bacteria. Mutualistic soil bacteria such as nitrogen fixers may be improved in numbers by growing legumes so as to host these nitrogen fixing bacteria.

3.6.2 Discouraging Harmful Soil Bacteria

In some situations, dangerous bacteria in the soil should be discouraged using practices that help reduce their population, reproduction or even their lifestyles in the soils. The main reason for discouraging harmful bacteria in the soils is mainly because they have devastating impacts on plants and soil in equal measure.

- The first method to control harmful soil bacteria from growing and multiplying in the soil involves providing unfavorable conditions. Anaerobic bacteria, for instance, may be discouraged by providing soils with good drainage thus oxygen will enter the soil and stops them from anaerobic respiration or fermentation. On the other hand, oxidizers may be controlled by reducing the amount of oxygen in the soil which is useful in oxidation. These conditions will stop aerobic soil bacteria from metabolism and oxidation of compounds in the soil.
- Chemicals may also be used to discourage pathogenic disease-causing bacteria. Chemicals such as pesticides inhibit the growth and reproduction of such harmful bacteria in the soil and consequently reducing their population and effects on plants and soil.
- Thirdly, soil bacteria may be discouraged by reducing the supply of organic matter to farm soils. This can be achieved by removing all plant residues and also through sanitization of farm soils. This way bacteria lack food for growth thus they decrease in number.
- Finally, soil bacteria may be discouraged by introducing antagonistic soil microorganisms that act to push out harmful soil bacteria from soils. Fungi are best examples of microorganisms that can be used to control the growth of harmful bacteria in the soil.

Chapter 4 Soil-Borne Fungi

CONTENTS

4.1 INTRODUCTION

Soil fungi are plant-like microscopic cells which in very long thread-like structures making a mass known as mycelium. Once this mycelium colonizes roots, they will absorb its nutrients. They also take surface organic matter and the surrounding soil. These soil fungi will produce spores for reproduction. Fungi structures are different but their action is always the same. Yeast, for instance, is a single-celled type of fungi. It because of this fact that fungi are not plant- specific when compared to soil bacteria like rhizobia.

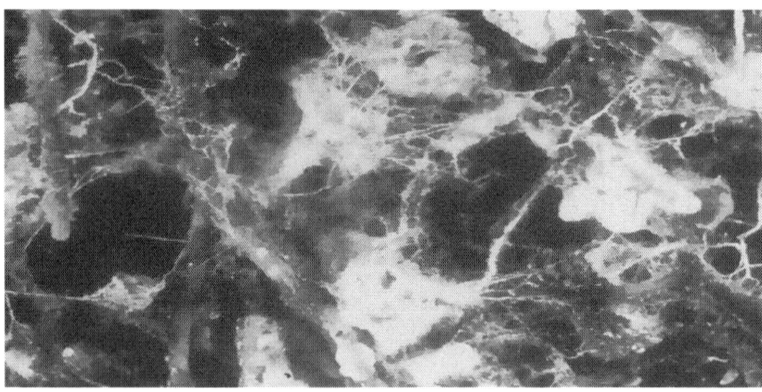

Figure 4.1: Fungi hyphae in the soil.

Source: https://www.nrcs.usda.gov/Internet/FSE_MEDIA/nrcs142p2_050027.jpg

It is true that while fungi may be useful in the decomposition of organic matter, it is similarly important to understand that other fungi may cause biological and chemical breakdown of pesticides into dangerous compounds. Moreover, the growth of plants from the affected soils is greatly affected by fungus activity. Seedling affected by soil fungus will grow much slower than normal seeds (Sarmiento et al, 2017).Another negative effect of soil fungus is causing crop diseases. Black root rotting on beans is largely associated with soil fungi. The study of soil fungi is of great importance in soil science. Soil fungi are very crucial in the study of soil stabilization. Soil fungi, therefore, are closely influential in soil properties.

In conclusion, it is important to note that most fungi species must have occurred in soil during some stage in their life cycle. Soil scientists should do more studies to help in better understanding of soil fungi so as to unravel myths and perceptions relating to soil fungi.

4.2 SOIL CONDITIONS WHICH FAVORS FUNGAL GROTH

Several soil factors and properties play a very important role in the growth of soil fungi.

- First, the soil temperature is important in the growth and reproduction of soil fungi. Soil scientist believes fungi grow and reproduce more in temperate soils especially temperatures 25-30°C. Low temperatures result in poor growth of soil fungi. However, very high temperatures will kill most soil fungi.
- Secondly, soil moisture determines the growth of soil fungi in soil. Generally, high moisture conditions of soils result in better growth of microbial species particularly soil fungi. Many species of soil fungi are highly inhibited by low soil moisture.
- Soil pH levels are also associated with fungi growth in soil. Recent studies reveal that higher pH values reduce the growth of soil fungi. Lower pH values were found by soil scientists as the best conditions for the growth of soil fungi.
- Soil substrates such as glucose and gelatin are crucial in the rise of soil fungi. The availability of such substrates in soil was thus found as important ingredients that enhanced the growth of soil fungi.
- Finally, the growth of soil fungi is greatly influenced by toxicity of the soil. It is found that toxic soils particularly those treated with antifungal antibiotics were found to discourage their growth. Soil fungi were therefore found to rapidly grow and reproduce in less toxic soil conditions.

In conclusion, it is important to note that soil conditions that encourage soil fungi are closely interdependent and are also influenced by other external factors.

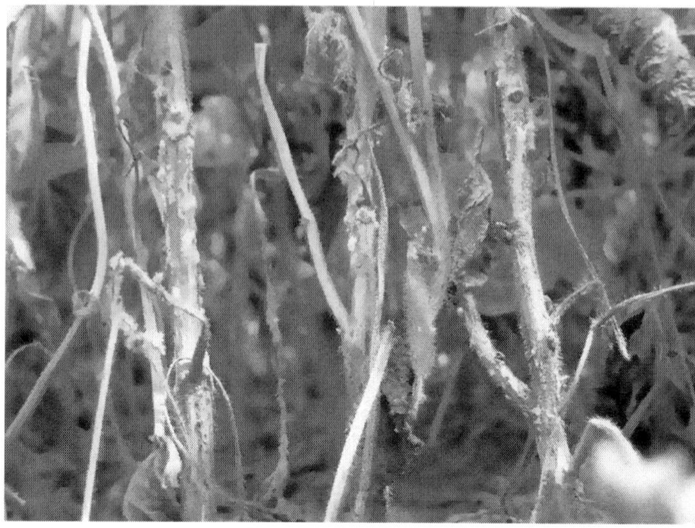

Figure 4.2: White mold on soybean.

Source:https://cropprotectionnetwork.org/resources/articles/diseases/white-mold-of-soybean

4.3 CLASSIFICATION OF SOIL FUNGI

4.3.1 Zygomycetes

Zygomycetes are saprophytic fungi found in soil. They are also known as conjugated fungi. Sometimes they also occur in decaying matter, skins and hair coats. This group is believed to be the smallest in the kingdom fungi. Zygomycetes belong to phylum Zygomycota (Sarmiento et al, 2017).

Most zygomycetes rapidly grow on surfaces of a host such as bread and fruits. Many zygomycetes are saprobes living off decayed or decaying matter. Some zygomycetes are parasitic to plants, small animals and some insects. Therefore, zygomycetes are commonly found in torrential habitats. The reproduction of Zygomycota is mainly asexual through production of sporangiospores. However, they may reproduce sexually if favorable conditions exist. Sexual reproduction of Zygomycota occurs when two opposing mating strains are fused so as to share their genetic content and consequently creating zygospores.

Zygospores will finally undergo meiosis when the right conditions are available. This leads to the production of haploid spores which develop into new zygomycetes organisms.

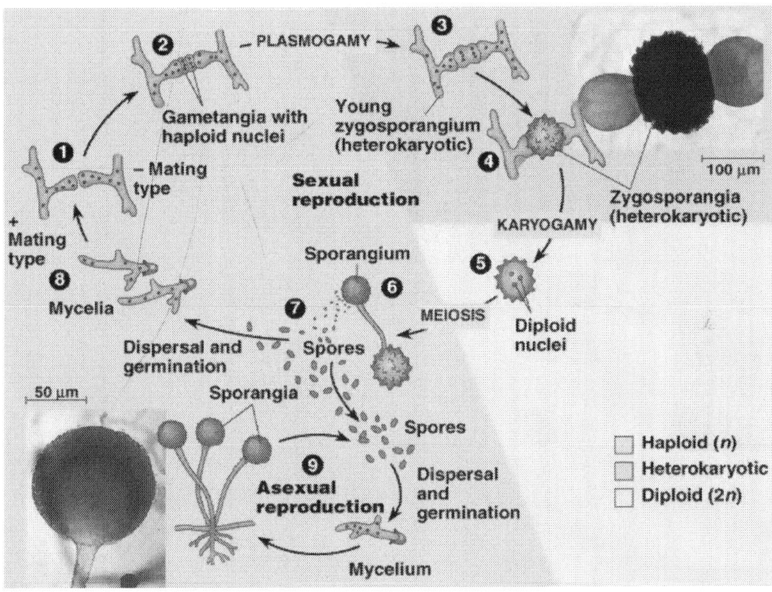

Figure 4.3: Illustration of Zygomycete life cycle.

Source: https://microbewiki.kenyon.edu/index.php/Zygomycota

The roles of these fungi are varied thus studying them is important. Zygomycetes, for instance, assist in decomposition of simple sugars in the soils and other torrential habitats. Some are plant mutualists while other species are pathogens to plants and animals (Sarmiento et al, 2017).

The negative effects caused by Zygomycota cannot be overlooked. They play a role in destroying fruits with high sugar content. They also destroy bread. Some fungi in this group cause dangerous human infections.

Figure 4.4: Zygomycota growing on fruits.

Source: https://www.wisegeek.com/what-is-zygomycetes.htm

In conclusion, most Zygomycota are not readily encountered by men because they lack common names of identification. Studies on this unknown Zygomycota should be carried out extensively.

4.3.2 Ascomycetes

This fungus is also known as the sac fungi. Many fungi known by man belong to this phylum Ascomycota. Ascomycetes are characterized by a sac-like structure known as ascus containing haploid ascospores (De la Rosa, Miller & Knicker, 2018).Ascomycetes mainly reproduces asexually. Ascomycota produces conidiophores that lead to new organisms of Ascomycota. On the other hand, ascomycetes may reproduce sexually by developing hyphae. The combination of male and female strains during fertilization produces special ascogeneoushyphae. The result is rise of new mycelia.

The importance of ascomycetes is crucial both to man and the environment. Yeast, for instance, is beneficial in baking and fermentation of wines. They are also used in delicacies while A.oryzae is useful in rice fermenting into sake (De la Rosa, Miller & Knicker, 2018).

Ascomycota may also have negative devastating effects on farm crops, animals and even humans. Some act as parasites such as fungal pneumonia which affects many AIDS patients. Moreover, ascomycetes in soils may destroy crops or even make yields unsuitable for human or livestock use.

This happens when Ascomycota in the soil or plants produce dangerous metabolites.

Ascomycota have specific characteristics despite being diverse. In terms of their morphology, most ascomycetes are unicellular. Yeast, for instance, is a unicellular Ascomycete. It is simple in structure and exists in unicellular form. Ascomycota are also characterized based on their habitats. Some Ascomycota are found in torrential habitats such as decomposing matter. a good example is Aleuria aurantia found in humus.Other species are found in marine environs while others are found in human hosts. Mycorrhizal species are commonly found in plant roots (De la Rosa, Miller & Knicker, 2018).

Figure 4.5: Ascomycota growing on dead material.

Source: http://bioexampreparation.blogspot.com/2017/10/ascomycota-structure-reproduction-economic-importance.html

4.3.3 Basidiomycetes

This fungus is also known as club fungi because of their fruiting club-like bodies. These fruiting bodies are known as basidia. These are swollen terminal cells.Basidia act as reproductive organs. They are common in mushrooms which can be seen in fields, shelves and gardens. Basidiomycota includes shelf fungi found in tree barks.They also consist of rust and smuts which act mainly as pathogens in plants. Some basidiomycetes are edible such as mushrooms while others may be toxic and consequently cause respiratory problems (De la Rosa, Miller & Knicker, 2018).

Mainly Basidiomycota members produce spores through sexual reproduction more than through asexual means. In this scenario basidium will carry spores known as basidiospores. This results in fusion of mating strains which produces diploid zygote which goes through meiosis so as to produce new organisms of Basidiomycota.

Basidiomycota have many distinct characteristics thatdistinguish them from other fungi. To begin with Basidiomycota are highly evolved in nature. In fact, their hyphae have a unique cellular structure characterized by single haploid nucleus. Moreover, many of these fungi group grows mainly as dikaryons until such a time when the circumstances make them produce fruit bodies like toadstools.Most Basidiomycota produce toadstool-shape like fruit bodies, with basidia lining spores. However, gasteromycetes very different fruit bodies along with basidia maturing inside the fruity bodies e.g. in puffballs which can be locally found in our day to day environment. Final it has been found that some Basidiomycota produces rings called fairy rings mainly during autumn. Most of this species of fungi produce green rings and will have no fundamental impact on grass (De la Rosa, Miller & Knicker, 2018).

4.3.4 Deuteromycota

They are commonly referred to as imperfect fungi. This is mainly because they have failed to be biologically or taxonomically classified. Another reason for their imperfections is their failure to demonstrate sexual reproduction. Though their asexual reproduction has been observed, imperfect fungi leave a lot to be studied by scientists particularly biologists. Most Deuteromycota, therefore, may be studied by deeply looking at their asexual reproduction. They normally reproduce via sporogenesis which leads to production of conidiophores. The asexual reproduction and the consequent production of spores by Deuteromycota have help in classification of these imperfect fungi.

- Hypomycetes - Do not have any fruiting bodies.
- Coelomycetes - Produce their spores in fruiting bodies.
- Agonomycetes- They lack any spores at all.

Most species of these fungi reproduce by conidia because they have lost the ability of sexual reproduction. Conidia are majorly produced from conidia pores resulting in new Deuteromycetes.This group of imperfect fungi is likely found in soil but may also be found in foods and air because most are airborne. They are also members of other phyla such as ascomycota. However, they cannot be morphologically classified because they are usually nonsexual in nature (De la Rosa, Miller & Knicker, 2018). It is because of this fact that further characterizations of these unique fungi have been made leading to classification as follow;

- Blastomycetes - Asexual phases of mycetous or dimorphic fungi.
- Hyphomycetes – Asexual conidial fungi.
- Coelomycetes – Asexual fungi thatproduce multiple hyphae called conidioma.
- Mycelia sterilla – Fungi of nonconidia which reproduce through fragmentation.

4.4 GENERAL FUNCTIONAL GROUPS OF SOIL FUNGI

Soil fungi are categorized into three functional groups. These are:

4.4.1 Decomposer Fungi

They form the first group of soil fungi. They are known as saprophytic fungi. Their main function is to convert organic matter into biomass, carbon 4 oxide and organic acids in the soil.Decomposers soil fungi are crucial in decomposing hardwood organic matter. Once decomposers consume organic matter, they make nutrients immobile thus retaining them in the soil. Moreover, the production of organic acids as a byproduct of decomposition results in the production of resistant organic matter that cannot be broken down easily. For instance, the decomposition of proteins and cellulose that is resistant to breakdown (Jenkins, 2005).

Some soil scientists describe soil decomposer fungi as soil magicians. It is this reason which is important to note that the same soil magicians decompose dead organic matter in phases which can be described as follows;

- Phase one mainly involves sugar fungi which are the first to attack compost and litter. They then take simple sugars while avoiding complex fiber. They also produce antibiotics to target bacteria. They can always stand high temperatures up to 75 degrees Celsius in the dead organic matter. Consequently, these sugar fungi will die when their food isdepleted.

- Phase two consists mostly of sac fungi, mushrooms and even brown rot fungi. At this stage, fungi decompose mostly woody organic matter. Cellulose, in particular, is decomposed at this stage. Fungi also at this stage specialize in what they decompose Amorphoteca will only decompose jet fuel.
- Phase three involves mushrooms producers and white-rot fungi. The first three mainly take place at cool temperatures. Here complex plant residues are broken down such as waxes tannins and lignin leaving white remnants hence the name white-rot fungi.

4.4.2 Pathogenic or Parasitic Fungi

This group includes well-known fungi in the soil such as Verticillium and Pythium. Their main function involves colonizing plants and decomposing living tissues which create weak malnourished and dead plants. Parasitic fungi are pathogenic in nature and are most dominant in soil (Jenkins, 2005). It is therefore because of this parasitic nature of these fungi that leads to devastating effects on crops. Devastating effects of parasitic fungi have been witnessed in forest where they spread diseases from plants to plants, for instance, powdery mildew which destroys chestnut trees.

Moreover, pathogenic fungi damage plants by causing rusts. Puccinia sparganiodes, for instance, needs many hosts in order to complete its life cycle thus causing a lot of crop damage. Another crucial pathogen is boil-smuts which are considered a delicacy by Mexicans but actually harms crops and reduce yields.

It is important to highlight that some soil fungi are also parasitic to humans. Once in the human body, pathogenic soil fungi cause infections and other complications. For instance, once inhaled pathogenic fungi spores will end up in the lungs causing complex lung infection (Jenkins, 2005). Additionally, it is notably useful for man to understand soil fungi and how they are transmitted. Most pathogenic fungi enter man's body via ingestion, inhalation or skin and wounds. Once in the body, parasitic fungi attack various parts of the body such as lungs skin nails and the mouth full stop, for instance, Candida ablicans affects newborn babies causing white patches in the mouth mucosa leading the discomfort and illness in newborns (Jenkins, 2005).

Parasitic fungi are also useful in other instances. Some pathogenic fungi are Acura useful to human in agriculture. They colonize roots and help plants build resilience to infections from other soil microorganisms or even from other dangerous fungi.

4.4.3 Mycorrhizal Fungi

They are also known as mutualist or mutualistic fungi. These soil fungi form a mutual association with vascular plants. In this case, they colonize plant roots either intracellular or extracellular. Commonly known plants that form mutualistic relationships with soil fungi include blueberry and Brazil nuts. It is believed that most mutualistic fungi form associations with 80% of plant species on earth (Jenkins, 2005).

Most soil scientists and biologists believe that mutualistic fungi evolved from the decomposer fungi but they lack ability to degrade complex organic matter like lignin. During mutual relationships with host plants, these fungi have access to carbohydrates. On their side,they help plants by assisting them to absorb phosphorus, water and other important mineral nutrients from the soil (Jenkins, 2005).

Mutualistic fungi can be categorized into different groups based on where they colonize in the plant. Ectomycorrhiza is a type of mutualistic fungi thatcross on the surface of most tree roots. On the other hand, endomycorrhiza is a mutualistic fungi type thatmajorly grows in root cells. They are commonly found in vegetable shrubs and grass.

The level of dependency between mutualistic fungi and plants vary from one group to another. Most trees and groups associate greatly with these soil fungi but some groups do not, for instance, spinach and mustard. Thesecrops and plants failed to interact and form any mutualistic or useful relationship with mycorrhizal fungi. Furthermore, a mycorrhizal fungus is likely affected by agricultural practices or conditions. Land with a lot of cohorts reduces mycorrhizal fungi growth and reproduction. The application of fungicides and constant land tillage also reduces mutualistic fungi in agricultural land. High concentrations of phosphorus and nitrogen in fertilizer applied tofarms contribute negatively tothe loss of mycorrhizal fungi (Jenkins, 2005).

4.5 BENEFITS OF SOIL FUNGI

Soil fungi play a very important role in the soil, plants and even to the environment. The following are major roles of soil fungi;

4.5.1 Soil Structure

A soil fungus plays a critical role in maintaining soil stabilization. Soil fungi produce hyphae which helps in binding soil particles. This way the soil will be more stable and retain or hold more water due to increased infiltration and drainage (Jenkins, 2005).

4.5.2 Manufacture of Antibiotics

Soil fungi such as decomposers are very useful in the production of antibiotics for medical purposes. The decomposition of dead organic matter by fungi results in the production of products such as carbon iv oxide and antibiotics. This process has been used biomedical engineers to help in the production of antibiotics which are consequently used to control diseases (Jenkins, 2005).

4.5.3 Absorption of Water

Soil fungi particularly mutualistic soil fungi boost the ability of plants and their roots to absorb water from the soil. Most farmers in semi-arid areas cultivate mycorrhizal fungi so as to add them to their farms to assist crops in water absorption.

4.5.4 Mineral Uptake

Mycorrhizal fungi are very useful to plants because they help them to take up minerals from the soil. Once these mutualistic fungi form association with plant roots they help the host plant to take up minerals such as phosphorus, nitrogen and other important nutrients from the soil. At the same time,these fungi will have continuous access to carbohydrates from the plants (Jenkins, 2005).

4.5.5 Decompose Organic Matter

Saprophytic soil fungi help in the decomposition of dead organic matter in the soil or on the soil surface. This breakdown process results in the production of useful soil nutrients and one is harder to the soil they will help improve the properties of the soil such as soil pH and soil fertility among other soil properties.

4.5.6 Uptake of Fertilizer

The ability of plants to absorb fertilizer is also promoted by soil fungi. Mycorrhizal fungi, in particular, help plants to absorb fertilizers because of their high absorption ability compared to plant roots. Soil fungi are used by farmers to assist their crops absorb fertilizer much faster.

4.5.7 Resilience and Defense Mechanism

Mutualistic and parasitic fungi sometimes protect plants from other dangerous fungi which may cause plant diseases. For instance, when mycorrhizal fungi colonize plant roots, they will protect the roots from being in contact with disease-causing organisms such as bacteria and other dangerous fungi or viruses and therefore providing plants with an effective defense mechanism against infection (Jenkins, 2005).

4.5.8 Nitrogen Fixation and Absorption

It is mainly because of soil fungi that most plants are able to get nitrogen from the soil. Mycorrhizal fungi are normally introduced to crops which have poor nitrogen absorption ability. The introduced soil fungi will assist these crops to absorb nutrients alongside the much-needed nitrogen (Jenkins, 2005).

4.5.9 Delicacies

Some fungi in the soil are crucial to man because they are used as delicacies. A good example is edible mushrooms. These edible mushrooms are believed to have very high mineral content which isvery crucial to man.

In conclusion, soil fungi have numerous roles and benefits both to man and other microorganisms in the soil.The study of soil fungi by soil scientists is, therefore, a very crucial subject and therefore research should be undertaken on this important microorganism.

4.6 COMMON DISEASES CAUSED BY SOIL FUNGI

Soil fungi despite their importance both to plants and soil are associated with various diseases. These fungal diseases are discussed below as follows.

4.6.1 Fusarium Wilt

This is a common plant disease caused by soil fungi known as Fusarium oxysporum. This fungus lives in the soil. It is a saprophytic fungus and stays in temperatures above 24 degrees Celsius.Plants once infected by this fungus will be stunted characterized by pale green leaves which turn golden yellow and finally wilt. It is also characterized by dark steaks root tissues and lower stems. Infected rootsdecay while germinating seeds die once infected (De la Rosa, Miller & Knicker, 2018).

Many plants are infected by this fungal disease. Examples of crops include legumes, melons, tomatoes, and sweet potatoes. Cases of this fungal infection havealso been detected in bananas.

The spread of this fungal disease maybe controlled and managed through the use of clean sheets. Another measure includes removing infected dead plants remains from the farm. Sometimes this fungal disease is host specific and the farm practices may not be able to control it thus the use of fungicides is highly recommended (De la Rosa, Miller &Knicker, 2018).This fungal disease not only destroys crops and trees but also destroy other saprophytic fungi living near or in the dying plants.

The spread of this fungal disease is lately through water splashing over short distances or planting and other farm tools which may spread the disease to the nearby environment. Long-distance migration and spread of the disease are mainly caused by infected plant transplants and seeds (De la Rosa, Miller &Knicker, 2018).

It is important to know that these fungi affect groups worldwide. For instance, this fungal disease causes wilting in bananas in Africa America and Asia. The growth of this fungus which causes wilting is mainly favored by warm soils in various temperatures depending on the infected plant.

4.6.2 Brown Patch

This is a common plant disease caused by a fungi pathogen called Rhizoctonia solani that affects all cool-season turf grass species.It is a foliar disease that does not affect crowns or roots. It causes moderate to severe outbreaks on high maintenance bent grass and bluegrass. This results in poor quality turf that is predisposed to algae and moss infestation.

Mild brown patch destroys golf grass appearance. Taller turf grass in athletic fields and landscape sustain will, however, sustain brown patch infections.

Major symptoms on plants affected by this fungal disease include:

- The brown patch causes circular olive-green stains of 4 to 12 inches
- The leaf blades within the patches turn brown after infection but the grey-white band is evident at the perimeter of the patches.
- Smoke rings are caused by mycelium and water-soaked in leaves. Smoke rings are commonly found in turfgrass but less evident. Individual lesions on leaf blades with brown margins occur in affected grass but very common in tall grass.

Thepathogens are active during hot and humid periods and very severe when nitrogen fertility is excess. The disease spreads through the expansion of mycelium overleaf blades and through poor farm practices in the farm due to fungi producing no spores.

This fungal disease can be controlled through cultural practices thatsuppress the fungi and also chemical control. For instance, avoid excess nitrogen fertilizer during summer to reduce the disease. Use fungicides and improve air circulation and propagation to reduce long dews so as to suppress outbreaks (Jenkins, 2005).

4.6.3 Downy Mildew

Downy mildew is a fungal disease of the foliage. It is caused by soil fungus such as Oomycetes. Mainly spread from plant to plant through airborne spores. It is very common in wet weather conditions because infections are promoted by prolonged wet leaf surfaces. This fungal disease affects plants such as carrots, foxgloves, grapevines, onions tobacco and spinach among others. The disease has few ranges such as species attacking branches while other species of the same fungi affect the roots (Jenkins, 2005).

This fungal disease has complex symptoms but the recognizable are commonly the following;

- Discolored blotches on leaves surfaces ranging from pale green purple yellow or brown based on the infected plants
- Blotches may have straight edges bordering near leaves veins
- Mold growth on lower leaf surfaces. They are usually green white or purple growths which can be seen in crops but not discoverable in foxgloves.
- Seriously infected leaves turn brown or yellow and fall off prematurely.

- Sometimes infected plants have stunted growth and poor health hence instant death

This fungal disease can be controlled by collecting and disposing of infected plants by burning once noticed. Remove and destroy all infected plants residue to reduce spreading to the surrounding. Furthermore, control weeds and ensure proper air circulation in the soil. Avoid cases of long periods of leaf wetness, for instance, do not irrigate in the evening so as to reduce leave wetness overnight.Finally practice crop rotation in vegetables to reduce infections only from spores living in the soil. Similarly cultivate downy mildew resistant crops especially onions if the place is prone to downy mildew.

4.6.4 Early Blight

This is a fungal disease mainly caused by soil fungi known as Alternaria solani. This fungal disease is popular in tomato and potato plants and causes devastating effects if not managed as soon as possible.In moist and cool conditions, the fungi produce spores in large numbers which are consequently spread by wind or splashing water thus causing infections and spreading them. Spores penetrate infected plant tissues causing devastating effects resulting in lesions (Magdalena et al, 2018).

A brown lesson develops on leaves which consequently become dry and papery developing concentric rings in the tissues. They also cause dark slightly sunk regions on the stems which are later enlarged when the infection gets serious. In the case of tubers, they develop dark lesions surrounded by purple margins. During storage, tubers are shriveled. Infected fruits have rings and a black mass of fungi growth covering the lesions.

Prevention and management of this serious fungal disease can be achieved by majorly true chemical control by using fungicides. The timely application of fungicides is therefore recommended. Moreover, there is need to alternate fungicides to ensure the pathogen does not develop resistance to them.Majorly cultural practices help reduce this fungal disease i.e. reduces splashing of water irrigate during daylight to allow crop services to dry. Furthermore, use healthy seeds and eliminate infected group residues after harvest. Finally allow proper sunlight penetration to the farm and always use clean farm equipment to avoid and reduce spread of this fungus both over long times and distances.

4.6.5 Late Blight

This is a fungal disease affecting mainly potatoes and tomatoes. It has several devastating effects on potatoes and tomatoes. If not checked immediately it will destroy completely the whole parts of the crop such as leaves stem and fruits. In some places, this fungal disease has been found to affect tubers.Major symptoms are noticeable once this fungal disease affects leaves branches and fruits. Leaves can turn pale green with brown spots and purple margins appearing on leaves.Spots will be growing larger as the infection progresses until the leaves die off. White grey mold grows on the leaf margins during moist conditions. Affected parts will look dry during dry weather but stems develop grey watery brown lesions (Sarmiento et al, 2017).

In tomato fruits, a watery grey spot isseen on the other part and turns brown and bumpy. A white fungal growth can be seen on the affected fruit mainly in moist conditions. In potatoes, infected tubers are watery. The above symptoms usually make the fruits or the tubers to develop a bad smell due to constant breakdown of plant tissue by the fungi.Generally, late blight effects on plant stages and all the sections of the plants. The spread of this disease is usually associated with fungal spores which can be caused by water splashes or wind. In the case of tuber crops any remaining plant tissues or remnants in the soil will encourage future infections by this fungal disease.Furthermore, the disease can be controlled by sanitizing farms, crop rotation, discourage overhead irrigation and removing all infected plant remains. Additionally, plant satisfied seeds and do not harvest tuber during wet conditions (Jenkins, 2005).

4.6.6 Leaf Curl

This is a fungal disease caused by fungi called Taprina deformans. The disease naturally affects leaves and then shoots. It is also common in fruits causing reduced production.During winter the fungus affects buds and fallen infected leaves but during spring, the fungi spores are spread to developing leaves. Wet conditions make the spores to grow in young leaves where they initiate infection. When the temperature increases this fungal disease is slowed. Therefore, this fungal disease is common during spring (Jenkins, 2005).

The major symptom this fungal disease is the curling of leaves with reddish areas which later results in distorted leaves. At later stages of the disease infected parts turn yellowish-grey covered with fungal spores.

Consequently, infected leaves turn brown or yellow while falling prematurely. Though the affected shoots develop new leaves wet conditions will occasion the infection of these shoots leading to their thickening and death.

This disease is manageable byplanting healthy seeds and avoiding excess fertilization. Mostly, healthy plants cannot lose leaves easily. Secondly, collect and dump infected leaves to help to reducethe spread of the fungal disease to other areas on the farm or the environment. Sanitize soils and clean farm equipment to reduce the spread of this fungal disease. Finally, it is important to practice thinning of fruit crops to help maintain plant vigor. Because this disease is common during spring, timely application of fungicides during spring will help prevent and control the disease. It is also important to practice crop rotation especially growing resistant varieties (Jenkins, 2005).

4.6.7 Powdery Mildew

Powdery mildew is a fungal disease thataffects plants by taking away nutrients making the affected plants weak. This fungal disease is caused by fungi of different species thatare plant-specific.This fungal disease is common in areas with high relative humidity. The spread of this disease is inhibited by cooler and rainy environments which discourage disease. Generally, the fungus that causes it slows down the growth of plants and may kill plants completely in severe infections (Jenkins, 2005).

The common characteristics of powdery mildew include plants looking as if dusted with flour. At first, the fungus causes white spots on the leaves,fruits and stems. They make leaves twist or severely destroy them completely. Leaf buds are sometimes injured.

This disease can be controlled by removing residues and destroying them through burning. Do not attempt to compost any plant residues of affected crops or plants because the fungus will spread by wind or water splashes. Chemical control is useful in controlling the fungal powdery mildew, especially in severe cases through the application of fungicides such as potassium bicarbonate.

Cultural control includes using plant varieties thatare resistant to powdery mildew. Moreover, discourage overhead irrigation to help reduce humidity. Furthermore, discourage watering during evening hours because this will not give plants time to dry and therefore you would have promoted relative humidity.Finally, pruning overgrown plant parts will help in sunlight penetration and air circulation thus reducing relative humidity around the plants consequently discouraging poultry mildew causing fungi (Jenkins, 2005).

4.6.8 Rust

It is a fungal disease that affects woody and herbaceous plants. Generally, the pathogen does not kill the plant. Rust will make the plants have poor health and very low production. This fungal disease caused by parasitic soil fungi affects plants in moist conditions. Fungi spores are transferred mainly by wind and splashing water from infected plants to the surrounding. Wet surfaces also encourage fungal infection of this disease.

Rust disease comes in different forms and affects a very wide range of plants. It is characterized by rust-like fungi. The main symptoms include white spots or yellow spots on the upper leave surface. Radish or orange swellings are seen on the lower side of the leaf. This is normally caused by growing fungal spores leading to leave distortion.

Rust disease is not easily treatable but can be controlled using the following; First, remove all infected plant residue. For fruits uproot infected plants then plant resistant varieties.Control weeds and remove debris between plants to stop the spread of the fungal disease to other parts of the farms or land. Reduce water splashes onto the leaves because this will encourage the spreading of the disease. The fungal disease may also be prevented by chemical control. Dust plants with sulfur early in the season to prevent infections. This will also keep infections from spreading to other healthy plants. Moreover, use a wide range of fungicides to treat infected plants and to prevent spreading.Proper spacing of crops ensures good air circulation and reduces rust spreading. Finally. avoid prolonged watering over leaf surfaces to reduce moisture around the plant (Jenkins, 2005).

4.6.9 Root Rot

This is a fungal disease thatattacks roots of plants particularly those growing in wet soils. This fungal disease mainly spread through the soil and majorly controlled by removing plant residues and destroying them completely.The disease is caused by soil fungi Rhizoctonia or Fusarium soil fungi which are commonly found in wet soil and therefore spreads from one part of the farm or land to another. Generally, the disease kills plants by reducing growth while suppressing the quality of the plant (Jenkins, 2005).

Major symptoms of root rot include; infected plants will grow slowly as compared to healthy plants. They will, therefore, look less vigorous with less production characterized by small leaves and flowers.Sometimes old leaves turn yellow and fall prematurely and leading to delayed flowering of plants because of the severe rotting of the roots. Rotten roots will look brown or black.

When plants are removed from the soil, the outer layer of the roots will strip off. This fungal disease is mainly promoted by over-fertilization, lack of adequate water and exposing roots to cool temperatures.This fungal disease can be controlled chemically through the use of fungicides however they are not easily killed by chemicals thus require repeated application of fungicides.Traditionalcontrol of this disease involves the sanitization of the farm and avoiding transplants from one part of the farm to another. Finally reduce root exposure to wet conditions and cool temperatures (Jenkins, 2005).

4.7 ECOLOGICAL PLANT- FUNGI INTERACTIONS

It is interestingly to note that the relationship between plants and soil fungi is not readily seen by many, however, plants and soil fungi build and maintain interactions underground. This relationship can be witnessed in roots of plants colonized by soil fungi. Other soil fungi decompose dead plants and woody matter.

Mycorrhizal fungi are an important soil fungus which demonstrates and have been found to form mutual relationship with plants in various ecosystems. A case scenario is in crops thatform relations with soil fungi. The soil fungi will provide the crops with various mineral nutrients such as phosphorus while the fungi will have constant access to carbohydrates from the plants. Many families of forest trees access minerals nutrients through fungi (Magdalena et al, 2018).

In desert ecosystems,plants will improve their water absorption by establishing mutual is the relationships with soil fungi thus accessthe water by using their relationship with soil fungi which have high retention and absorption.

Figure 4.6: Decomposer fungi relating with plants mutually.

Source: https://phys.org/news/2014-07-plant-fungal-interactions.html

The decomposer fungi speed up the decomposition of dead organic and woody matter. They do this by breaking down organic compounds leading to the production of carbon dioxide antibiotics and nutrients that can be absorbed by plants.In the case of parasitic fungi relations, they are usually devastating in any ecosystem. This is purely because plants will not benefit from a soil fungus. Some plants, therefore, may develop defense mechanisms to discourage parasitic fungi associations. Fungicides are also used to discourage such associations which may affect plant health and production (Magdalena et al, 2018).

Useful soil fungi have evolved in various forms which allow them to establish an association with plants. Therefore, despite plant defense mechanism some soil fungi will colonize them. They will then establish obligated relations through patterns of colonialism-basedfeeding styles and the plant's nutrients sources.Based on evolution, soil fungi are believed to be older and their interaction with plants is nearly same as higher plants. It is therefore because of this reason that many biologists believe that colonization of land by plants was really aided by soil fungi. It is therefore important to note that most plants on the planet earth host some endophytes soil fungi for symbiotic purposes. This is because they do not show symptoms of growing inside the plant tissues and secondly many plants relate with mycorrhizae soil fungi for effective nutrient uptake.Furthermore, in the movement of plants from aquatic ecosystem, it was mainly because plants wanted symbiotic relationships with soil fungi. In these case plants were aided to get mineral nutrients (Magdalena et al, 2018).

In conclusion, the plant-fungi relationship is a complex process thatmust be studied basing on several interrelated factors not only limited to the type and function of soil fungi but also soil properties and environmental factors present in various ecological systems all over the world. This studywill give more clear explanations on this concept which baffles many soil scientists and biologists in equal measure.

Chapter 5

Soil-Borne Parasitic Worms

CONTENTS

5.1 INTRODUCTION

A worm is an elongated, naked, soft-bodied animal that mostly depends on other animals and human bodies for shelter and food. There are numerous types of worms, depending on the kind of shape, size, and effect on the host's body. A variety of worms are parasites. This is because they depend on other organisms for their survival. Some worms are transmitted through the soil. This type of worms is referred to as soil-transmitted helminths. Soil-transmitted helminths are worms that infect humans and are spread through contaminated land. Helminths refer to parasitic infections. For instance, Ascaris lumbricoides (Ascaris), Trichuris trichiura (whipworm), and Anclostoma duodenale (hookworms) are soil-transmitted helminths. There are other several ways through which worms can be transmitted to human or animal bodies. Some are transmitted through water. For example, Worms like guinea worm and causative agents of giardiasis, cryptosporidiosis, and amebiasis are transmitted through water.

When these worms get in one's body, they cause infection and may cause death if not addressed as early as possible. For example, when one drinks water contaminated with parasite Dracunculusmedinensis, which causes guinea worm, he or she is likely to get a guinea worm disease, which is very infectious. When this worm is not eradicated in time, it can get more powerful and, therefore, can lead to death. Note that not all worms which are infectious, other worms like garden worms such as earthworms, flatworms, and nematodes are essential to the soil and plants.

5.2 WORMS FOUND IN THE SOIL

There are three types of worms found in the soil, namely:

- Earthworms
- Flatworms
- Nematodes

5.2.1 Earthworms

An earthworm is a tube-shaped segmented body soil living worm. It, therefore, means it feeds on dead and organic matters in the soil. Its digestion system runs through the body. From the body characteristics, the earthworm is under phylum Annelida. Earthworms have circular and longitudinal muscles on every segment, which allow them to move. They have a fluid-filled coelom chamber, which helps them to maintain their body structure. This is because they don't have both exoskeleton and endoskeleton like the other invertebrates. The fluid that keeps the body structure of earthworms is known as a hydrostatic skeleton. Every individual earthworm has both male and female sex organs. Organisms with such kind of characters are referred to as hermaphrodites. The fact that earthworms do not have eyes means that they have a unique mechanism which helps them in movement. They have photoreceptors which assist then in the photosensitivity. These earthworms' photoreceptor cells are known as light cells of Hess. The light cell of Hess is distributed in the epidermis of the earthworm. They are highly distributed at the side s and back of the worm. Like other living organisms, earthworms do excrete waste and toxic products from their bodies. Every earthworm has nephridia. Every segment except only the first three and last three segments has a pair of nephridia. The waste products in the first three and last three segments are propelled to the segments with nephridia by the movement of cilia, which are attached to microvilli in the segments (Hotez et al, 2006).

Figure 5.1: An earthworm.

Source: https://www.newsoptimist.ca/opinion/columnists/earthworms

5.2.2 Flatworms

A flatworm is a worm with no segments. Flatworms are one of the soil-transmitted helminths. They do not use any body cavity and specialized circulation and excretion systems. Some of the flatworms are hermaphrodites meaning every individual has got both male and female sex organs. Therefore, flatworms require a large surface to ensure the diffusion of nutrients and oxygen in the body. These characteristic places the flatworm in the phylum Platyhelminthes. Like other organisms under phylum, flatworms are multicellular. They, therefore, have a complex structure where tissues are organized into organs, which reorganize into an organ system. For instance, several organs in flatworm organize themselves to form an organ system such as the digestive system.

Although flatworms are free-living and not destructive, some are parasites. Some flatworms such as tapeworm and fluke parasitize in human beings, animals, or both. At the larval stage, these flatworms can cause severe damage to the host. Apart from affecting the host, it may lead to an infection to the consumers of the already infected host. For instance, when a sheep if infected by the tapeworm and a human being consumes the mutton, the human may be affected by the tapeworm. These flatworms carry with them some parasites known as rat lungworm that causes some meningitis to human beings. Other people are allergic to the toxic substances thatare secreted by the flatworms (Hotez et al, 2006).

5.2.3 Nematodes

Nematodes are structurally simple organisms that are tubular, that meaning has openings at both ends. i.e., mouth and anus. Nematodes are also referred to as roundworms. They are classified under phylum Nematoda. They have digestive, nervous, excretory, and reproductive systems. Besides, they have no specified circulatory or respiratory systems. Nematodes are on the most numerous multicellular animals on earth. Nematodes are microscopic, and one may not be able to see them unless they use a microscope. With a hand full of soil, there are thousands of nematodes. Nematodes feed on bacteria and fungi. They can also feed on other nematodes. Due to this, most of the species encountered are not well understood biologically.

Nematodes are naturally free-living. Even though most of them are not harmful, some nematodes cause diseases in both human beings and animals. Nematode Wuchereriabancrofti, for instance, is spread by mosquito bites. It damages the lymphatic system and produces large, debilitating swellings in different parts of the body. These swellings are referred to as elephantiasis. Other nematodes invent the human and animal stomach leading then to insufficient absorb nutrients. They can also destroy plants by boring some holes in the plant parts like the roots and stems. They can, in one way or the other, even cause a plant to dye. Most nematodes have very sharp mouths, and they can, therefore, suck the cell content in a plant when they bore the plant parts. Sucking of the cell content in the cell thus causes the death of the plant(Hotez et al, 2006).

5.3 IMPORTANCE OF EARTHWORMS

5.3.1 Earthworms help Improve the Nutrient Availability

Earthworms eat as they extract nutrients by burrowing of materials such as plants and debris or by dragging materials from plants into the burrows. Some bacteria and nutrients are produced when earthworms leave waste products through burrowing. There are three types of burrowing;

- Horizontal burrowing that contains shafts that are occasional into the soil.
- There is deep vertical burrowing which has a surface with many branch entries.
- The vertical burrows that are formed by the earthworms living in the soil.

The dead organic matter gets decomposed into nutrients that are important to the life of the plants. For example, earthworms decompose dead animals until they cannot be seen. The bodies of worm always decompose leading to soil rich in Nitrogen. The worm pile is said to be releasing four times more phosphorus than the soil can do. Earthworms usually leave the tunnels of their worm pile that are rich in nutrients hence leaves the soil suitable for plant growth (Hotez et al, 2006).

Figure 5.2: The earthworm cast.

*Source:*https://www.sciencelearn.org.nz/resources/9-earthworms-role-in-the-ecosystem

Soil worms show a significant impact on the soil microbial community and increment nitrogen mineralization in soil. Earthworms control nutrient supply.Through their tissues, earthworms control the supply of nutrients through burrowing primarily; aggregates and pores such as bio-structures are createdon the surface of the soil thereby influencing their physical properties, cycling of nutrients, and growth of plants. Biogenic systems are collections of organic-mineral materials.

Thestability and concentration of organic matter influence the physical properties of soil and the composition of organic matter. They also affect ecological soil processes that are vital in their "true domain" where resources and nutrients that are further exploited by populations of soil microorganisms are accumulated. The effect of earthworms on organic matter dynamics differs depending on the time and space scales. Endogenous earthworm behavior accelerates the initial turning of organic matter in the humid tropical environment as determinants of microbial activity by indirect effects on soil as microbial activity determinant. The content in the gut is often abundant in organic matter, nutrients, and water relative to bulk soil due to choosing organic particles foraging and can promote high levels of microbial activity (Patel, 2016).

5.3.2 Earthworms help Improve Soil Drainage

The channeling that is so extensive and the burrowing of the earthworms causes aeration of the soil and also makes the soil loose and therefore improving the soil drainage.Soils that contain earthworms drain quicker thansoil without earthworms and this is approximately ten times quicker. The infiltration of water is usually great in areas where earthworms are densely populated and with no-tillage. The tunnels of the earthworms normally are influenced by the rain, the irrigation followedand the soil gravity which acts as other materials pass way.As earthworms move through or into the soil, they enhance soil porosity. Some of them may make deep burrows that are permanent into the soil. After the death of the inhabitants, the burrows can still survive for long and can have a vital role in soil drainage especially when there are heavy rains. The porosity and the drainage of soil may be enhanced by horizontal burrows (Lemtiri et al, 2014).

The vertical burrowing is also important since it allows the infiltration of water to occur faster in areas where there is null till. Vertical burrowing forms large pore in earthworms' presence and when there is high water supply it leads to infiltration.

There is an increase in the capacity of water holding of soil when Earthworms enable the soil structure to aggregate by bringing or bonding the soil particles together.The channels for water and air are opened up by thetunnelingof the earthworms down into the soil. The worm leaves down and deep in the vertical burrows that are permanent which is approximately meters deep however they get their food from the surface of the soil. Root growth is facilitated by the breaking up of the tunnels and allowing deep penetration of roots to the ground hence more nutrients aremade available (Lemtiri et al, 2014).

5.3.3 Earthworms help Improve Soil Structure

Burrows are created in the soil as earthworm moves. These burrows create pores in the soil that enables the extraction of carbon dioxide from the soil and entry of oxygen into the soil. Vertical and horizontal burrows are made by different kinds of earthworms, in which a number of them may be so deep. Soil that has a lot of burrows does not flash flood (Lemtiri et al, 2014).

Figure 5.3: Earthworms and soil structure.

Source: https://www.farmersweekly.co.za/farm-basics/how-to-crop/keep-earthworms

The burrows are not only good for ventilation and drainage in soil, but they alsohave beneficial bacteria. These bacteria provide plant roots with rich nutrients.Crumb soil structure is also formed by the waste matter or worm cast. These waste matters help in improving soil fertility. Furthermore, earthworms enable the soil structure to aggregate by bringing or binding the soil particles together and hence increases the capacity of water holding of soil.

5.3.4 Earthworms help Improve Soil Productivity

Earthworms are a main component of faunain the soil communities in most biological systems and contain expansive biomass. The role of earthworm inthe growth of plants has been widely demonstrated, availability of nutrients in soils increases possibly. Earthworms play a critical part to increase the ripeness of soil in ways of assortment e.g.they bring the nutrients up from profound in the soil and store them on the surface of soil as cast, consequently the leaching of supplements is counteracted (Lemtiri et al, 2014).

Earthworms blend layers and join natural matter into the soil. This blending permits organic scattering through the soil hence making the nutrients held in it accessible to plants. Earthworms contribute to soil readiness when the soil structure is improved by it.

5.4 HOW EARTHWORMS CAN BE ENCOURAGED

Earthworm densities in the soil have proven to be a good indicator of exemplary soil health. They improve nutrient availability by feeding on plant debris. Organic and mineral components are concentrated in their digestive system releasing casts that have more nutrients than that in the soil. They also improve drainage through their extensive movement in the soil. All the burrows they create loosen the soil and allows for aeration, thus improving soil drainage. The casts they produce also help to maintain and improve the soil structure by holding the soil particles together aswater-stable aggregates which store moisture.

These are just but a few of the advantage's earthworm has to the soil. How to encourage earthworms in the soil is a topic that has not been discussed much. Below is an explanation of how we can improve soil health by increasing and encouraging earthworms in the soil.

5.4.1 Ensuring Soil pH is ($CaCl_2$) is above 4.5

Soil pH is paramount in soil health. It exerts an influence on how plants grow and whether the plants have enough nutrient availability. Soil pH is a measure whether the soil is acidic or alkaline. Earthworms have a preference of pH above 4.5. In acidic soil, earthworms get diseases or even die (Hotez et al, 2006).

A pH of 5.5-7.0 has proven to be the finest pH to promote plant growth. Macronutrient availability is increased in soils with a moderate alkaline state and micronutrients in an acidic state. The soil pH has an effect on beneficial microorganisms that are supposed to decompose organic matter. Acidic soils obstruct their functionality thus resulting in the over accumulation of organic matter and nutrients are therefore not readily released to the plants. Earthworms cannot survive in such conditions due to lack of nutrients which they also benefit from. Generally, limestone is used to raise the pH level. Burning plant residues or adding ashes is a source of Potassium, Calcium, Magnesium and Carbonates. Earthworms require Calcium in a continuous supply to survive. Liming materials such as $CaCO_3$ also increase soil pH. Lime is a certified organic product (Hotez et al, 2006).

Sulfur is used to lower the pH level. Elemental sulfur may be used to make strongly alkaline soil neutral or to the pH that is desired. When elemental sulfur is applied to soil, some microorganisms assists in the combination of S, O and H_20 (which takes three to six weeks) to create sulfuric acid. The pH is measured using Calcium Chloride.

5.4.2 Increasing Organic Matter

Earthworms feed on organic matter and soil. They mix dead leaves and other organic matter with the soil, making the litter easier to decompose by soil microorganisms. Soil organic matter is composed of living organisms. Some of these are earthworms, bacteria, fungi, algae, nematodes, plant debris and humus. Up to 15% of organic matter is leaves from plants or any other plant debris. Organisms in the soil assist in the decomposition of this matter in the right conditions, assisting in providing nutrients to the soil. Organisms create by-products, wastes and cell tissues. Compounds released as waste by one organism can be used as food by another (Hotez et al, 2006).

Soil organic matter can be maintained by proper farming methods such as crop rotations, use of cover crops, proper methods of cultivation and use of fertilizer.

Humus has many functions that improve soil health;

- It retains moisture and nutrients.
- It reduces soil erosion by helping compact the soil to the required

amount without taking all the soil spaces away.

- Due to microbial activity, it is slowly decayed thus releasing some nutrients such as Nitrogen to the soil.
- It buffers the effects of pesticides.

Some of the farming practices that provide food for earthworms are use of permanent pasture which continues to provide organic matter to the soil from dead leaves and roots that decay. Green manure crops are cultivated primarily for animal feed but can be used by leaving them on the surface of the soil or turning them into the soil to provide organic matter. Crop debrisis important in providing organic matter to the soil. Many farmers burn debrisbut this may kill earthworms and destroy the surface organic matter. It is better left alone to decompose. One can sow the debristo the crops that will help increase the number of earthworms as well (Hotez et al, 2006).

5.4.3 Reduce use of some Fertilizers and Fungicides

Despite the fact that fertilizers have certain advantages such as a rise in yield and is a source of many essential elements, some fertilizers may lower the soil pH making the soil acidic. Fertilizers such as ammonium sulfate are very acidifying. Earthworms get ill and may die in the acidic soil thus reducing their numbers. Fertilizers also have other disadvantages; they are slow to release soil nutrition, they kill birds, rats and other crops disrupting the pyramid and are harmful to the soil.

Herbicides are weed killers. When earthworms eat plant material with some herbicides, it may cause toxicity and make the earthworms ill. The change in the chemical environment may also draw away the earthworms due to the inhabitable environment. Especially if the environment is made acidic by the herbicides (Hotez et al, 2006).

Herbicides and fertilizers should be avoided in large amounts so as to avoid changing the chemical environment of the earthworms by acidifying the soil, or by coating the earthworms and making the sick or killing them.

5.4.4 Keeping Soil Moist

Worms need moisture to survive because they lose around 20% of their body weight each day in mucus and castings. To increase moisture availability in the soil, it is necessary to adopt moisture conservation techniques. The main reason to conserve soil moisture is to prevent the loss of water from the soil.

Water loss can be caused by transpiration or evaporation due to the sun's heat. Some are listed as below.

- Deep tillage which is when tillage is performed below the normal length using methods such as ripping or subsoiling so as to change the physical or chemical properties of the soil. This method is only used on a need to use basis as it can kill earthworms that are residing deep in the soil.
- Mulching is one of the simplest methods to conserve soil moisture. This is a layer of a protective material that is placed or poured on top of the soil. Mulch can either be organic such as straw-back or inorganic like stone bricks. This minimizes loss of water by evaporation and also provides valuable nutrients through decomposition. The advantages of mulching are; it protects the soil from erosion, reduces or prevents compaction from the impact of heavy rains, it prevents weed growth and it maintains an even soil temperature.
- Building ridges and furrows when troughs are created by a system of plowing. It was used in Europe during the middle age.
- Crop rotation is the art of growing different crops in a sequence on the same piece of land over a specified period of time helps improve soil structure, therefore, improving the soil's water holding capacity. For example, rotating deep-rooted and shallow-rooted crops making use of previously unused soil moisture, as plants draw water from different depth levels within the soil. Crop rotation may also improve soil fertility and help control pests and diseases.

5.4.5 Improving Drainage

Drainage gets rid of excess water in the soil thus reducing water saturation and allowing soils to retain their original and fertile state. Most of this water saturation occurs on the top layer of the soil, making it easier to drain. The ultimate purpose of draining the soil is to be able to ensure the soil can support plant growth. Worms need reasonably aerated soils. Water saturation can cause a deficit of Oxygen in the soil, due to the spaces in the soil being filled by water. Water is known to have a lower percentage of Oxygen than air and the worms cannot get enough oxygen which they need to thrive from the soil if the soil is saturated with water (Hotez et al, 2006).

There are different types of soils, all with different properties. Some soils with clay may have a harder time draining water due to their close-knit, small particles. This promotes flooding in such soils. To encourage earthworms, one must choose the appropriate soil to plant in.

One can deal with poor drainage by increasing the amount of organic matter into the soil, or by the building of raised beds on top of the wet saturated ground. To identify poor drainage, one need to look at the color of the leaves which may be yellow or pale green, the integrity of the stem can also be observed.

5.4.6 Reducing Soil Compaction

Compaction is when the soil particles are pushed down close together mechanically. Air is expelled from the soil and mass density is increased. Compacted soil prevents the proper movement of earthworms in the soil. Earthworms move in the soil creating burrows which is advantageous because it improves soil drainage. When soils are compacted, the abundance of earthworms is decreased, the number of macro-pores is lessened (less air), and water infiltration is reduced (Hotez et al, 2006).

There are ways to prevent compaction. One should avoid wheel traffic. This means avoiding cars, mobile machinery and heavy foot traffic on the land to be cultivated. Minimizing tractor weight, avoiding tillage of wet soils, using wider tires or dual tires, maintaining minimum tire inflation, avoiding oversized equipment and combining field operations are few approaches.

5.4.7 Reducing Cultivation

Plowing the soil causes a reduction in the earthworm density. Earthworms on the surface especially are killed from tillage processes. Shallow cultivation has however proven to reduce the effect on the worms. Shifting cultivation is a method that leads to over-cultivation of a certain piece of land. Shifting cultivation is a farming system where farmers move on from one place to another when the land becomes exhausted. This is mainly practiced by primitive tribes in the forested highlands of the Amazon Basin, Congo Basin, etc. The advantages are;

- It allows the fast growth of the crops.
- It is environmentally fine because it all done using organic material.
- It often helps in the control of weeds.

- Due to the movement, it can help control pests.
- It may act as a way to control soil-borne diseases.

The disadvantages are;

- It is not long term.
- It is not good for land that is used only for one type of crop.
- It does not produce enough food.
- It is troublesome to move around all the time.
- It is not cost-effective.

This method is very destructive to the environment. This method can cause deforestation, soil erosion, loss of biodiversity and even pollution of coastal waters from oil residue and raw sewage. This can be avoided by using less aggressive forms of cultivation. Cultivating at appropriate soil moisture content (especially when using aggressive implements) is vital. When one uses implements that mix soil throughout the cultivation depth, one should restrict the depth of working to 5cm or less (Hotez et al, 2006).

5.4.8 Protection from Climate Extremes

Earthworms cannot survive in drought and frost. They also cannot survive in dry sandy soils. They can only survive in well-aerated, moist soils. There are many ways to protect the plants from extreme weather. Using organic matter helps protect the soil from extreme climate. Mineral nutrition is important in crop production. Mulching can be used to guard the soil by encasing it and preventing drying and cracking caused by evaporation of water from the soil. It also reduces the force of rainwater, especially during heavy rains, thereby preventing soil compaction. Examples are grass, paddy straws, sawdust, and even sugarcane trash. One could also use irrigation methods during droughts. This is the process of supplying water by artificial means to agricultural fields for crop production. This protects the soil from drying up allowing earthworms to survive. There are four methods of irrigation. Surface irrigation, Sub-surface irrigation, Sprinkler irrigation and Drip irrigation. The selection of the method of irrigation is based on; the soil characteristics, topography of the area, the available water supply, the type of crops and its requirements (Hotez et al, 2006).

Another way to protect the soil from extreme climate is by using greenhouses. These are structures with walls are roofs made chiefly of transparent materials such as glass where plants requiring regulated climatic conditions are grown. This can be done by cutting up pasture sods from places with earthworm density and using them to grow the plants.

5.5 SPECIFIC PARASITIC SOIL-BORNE WORMS

5.5.1 Roundworms

A roundworm or nematode has a "tube" like body within a tube, with a fluid-filled cavity in between the tubes. Roundworm belongsa phylum in the Linnaean Taxonomic Hierarchy. These modifications include the addition of more taxa and the appliance of cladistics to decide the classification of an organism.

Most nematodes are parasites, such as Ascarids. They usually enjoy moist, warm areas where an infection can thrive. Free-living roundworms recycle decaying rotting organic matter in almost all environments. Most roundworms are free-living and carnivores. Roundworms could be either decomposers or consumers. Roundworms help decompose organic matter and there would be a lack of energy supply to plants and animals. Roundworms facilitate the recycling of vital nutrients from bacteria, fungi and decaying matter allowing them to go back to the soil (Hotez et al, 2006).

They are transmitted in the fecal-oral route: means they are transmitted by eating eggs of nematodes that are passed through feces. The eggs in the feces are passed in the environment. They require warm moist soils. This protects the eggs and larvae from desiccation for several weeks. Once humans infected, the worms cause symptoms like cough, shortness of breath, abdominal pain, nausea, diarrhea, blood in the stool, weight loss, fatigue, etc. Roundworms have adapted to almost every environment. They can be found living on, in, or around nearly every living organism in virtually every habitat. They can go into cellular hibernation when their habitat is less desirable. They can also adapt deep in the soil on the ground and feed on many plants.

5.5.2 Whipworms

This is a soil-transmitted helminth. It commonly occurs with Ascaris. It is the second most common intestinal worm, especially in urban disadvantaged communities. It has worldwide distribution. It is more common in Asia, less in Africa and South America and rare in the United States. It has a long narrow anterior part and a short thicker posterior part. It is pinkish-white in color. The eggs are barrel-shaped with polar hyaline mucoid plugs.

The whipworm inhabits the large intestine. After copulation, the female worm lays eggs, which are passed out with feces and deposited in the soil. In favorable conditions, the eggs develop and become embryonated in the soil within two to three weeks. If swallowed, the infective embryonated eggs go to the small intestine and undergo four larval stages to become adult worms. Poor hygiene is a major factor in the transmission of trichuriasis. Children pose the highest risk of transmitting this parasitic disease. Prevalence of whipworm infection is higher in schools where hygiene conditions (i.e. tap water, hand-washing soap) are lacking. Adults reside in the large intestines, cecum, and appendix of the host. In heavily infected individuals, the infection can be present in more distal segments of the GI tract such as the descending colon and the rectum. It is transmitted via the fecal-oral route- when passage of stool is done on the soil. Children playing in the fields could contaminate their hands and end up eating the embryonated ova. It is more common in less-developed countries (Hotez et al, 2006).

5.5.3 Hookworms

This is one of the most important parasitic diseases of humans. The number of humans infected is estimated at 1.2 billion with 50,000-60,000 deaths a year.There are two types: Ancylostoma duodenale and Necatoramericanas. Ancylostoma has a buccal cavity with 2 pairs of large ventral teeth while Necator has dorsal cutting plates.

Man becomes infected when the infective larval stage (L3) which is found in soil, penetrates the skin. The larva enters the blood vessels in the epidermis, reach the heart then lungs (heart-lung migration). After one week, the larva migrates from the lungs through the bronchial tree, reaches the throat then swallowed and enters the intestine. In the small intestine, the larvae attach to intestinal mucosa and feed on blood. The adult mate and produce more eggs.Some of the symptoms is ground itch (first sign due to penetration of the larva) which includes itching, redness and skin creeping rash. It has a lung phase, with coughing and blood-tinged sputum. There is an intestinal phase, with nausea, vomiting, loss of appetite, stomach pains, constipation, diarrhea, weight loss, etc. Hookworms cause microcytic hypochromic anemia. This is a complication and the hemoglobin level can drop to 5g/dl. This occurs in chronic infection or heavy infection, due to continuous blood loss from wounds made by the worms in the intestine.

Chapter 6

Soil-Borne Protozoa

CONTENTS

6.1 INTRODUCTION

Protozoa can be defined as a group of non-phototrophic, eukaryotic and unicellular microbes having no cell walls. This comprehensive group consists of more than 65,000 sub-species, which basically share similar characteristics. However, this class of microorganisms may be particularly complex and flexible. Different from typical bacteria, protozoa may have multiple intracellular organelles doing specific functions.

Some types of protozoa even have parts that are similar to mouths and gastrointestinal tracts. These attributes probably contradict everything you might have learned about microbes, that they're just a sack of cells and enzymes. Currently, there are around 50,000 different forms of protozoa ranging from solitary microbes to those existing as colonies, while others are parasites which live inside plants and animals.

The longevity of protozoa can be attributed to how they behave which is quite similar to animals. Even though they appear simple, they can travel on their own and some even exhibit predatory behavior. For instance, amoebahas been known to expertly hunt other one-celled microbes before assimilating and slowly digesting them (Sukhareva-Buell, 2003).

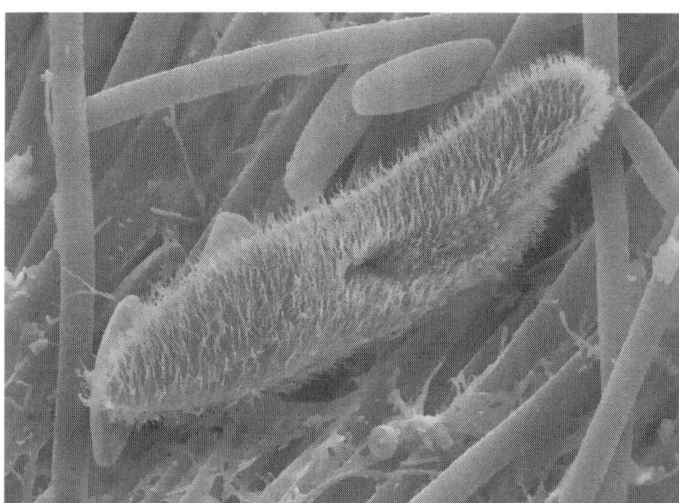

Figure 6.1: Microscopic image of protozoa showing its cylindrical shape and hairy body.

Source: https://microbiologyonline.org/about-microbiology/introducing-microbes/protozoa

6.1.1 Health Effects of Protozoa

Many protozoa species cause illnesses in both animals and people. Some, such as Plasmodium, are known to cause malaria, which devastates populations worldwide. Nevertheless, the greater majority of these organisms are totally harmless. Though, as it's the norm in microbiology, it's the hazardous ones that receive the most talk. Plus, with no vaccination available to safeguard against protozoan organisms, prevention still remains the most ideal defense while early treatment is also recommended. This parasite is capable of being transmitted through waste matter, should it find entry into the patient's digestive system or circulatory stream (Sukhareva-Buell, 2003).

6.2 THREE GROUPS OF PROTOZOA

6.2.1 Ciliates

Ciliates, also known as ciliophoran, are members of the Ciliophora protozoan family, which comprises of around 8,000 species. They are widely regarded as the most developed and multifaceted of protozoans. These organisms are single-celled and, at some point in their lifespan, may develop short hair-like protrusions called cilia, which are used for movement and gathering of food.

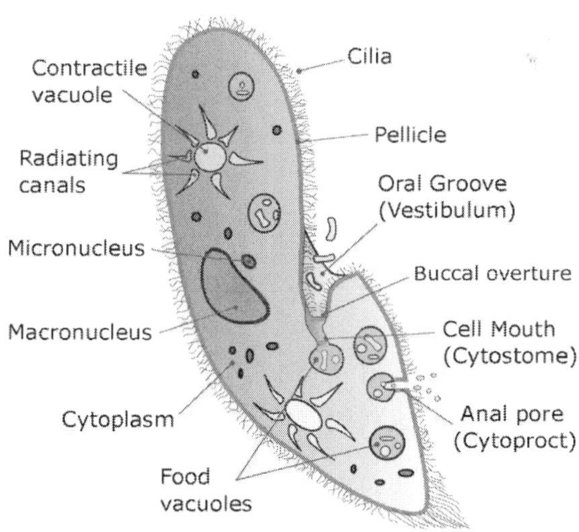

Figure 6.2: A ciliated protozoon with both micro and macro nucleus components.

Source: https://en.wikipedia.org/wiki/Ciliate

These cilia are commonly arranged in rows, referred to as kineties along with the cell covering, though they could also merge together around the cytostome or cell opening. This is seen in some species which form undulating membranes or membranelles, appearing as fan-shaped collections of cilia. Additionally, most ciliates comprise of an elastic pellicle and expandable vacuoles, while others have tiny organelles with stringor needle-like structures that can be released for anchorage, shielding, or for catching prey (Sukhareva-Buell, 2003).

Types of Ciliophora: The Ciliophora group comprises of well over 7000 microbes, it also comprises the largest species variety in most waste-water treatment facilities. Scientists have identified 4 unique types of Ciliophora which can be classified depending on their movement and cilia arrangement. They include the following:

- Spirotrichia: These have a flattened body and their movement cilia iscommonly found on the lower side. The cilia used for feeding are strongly developed and rotates clockwise from the cytosome. Some common examples of this species are Euplotes, Aspidisca and Stentor.
- Holotrichia: They are free-swimming microbes with cilia arranged neatly over their entire bodies. Typical strains are Litonotus and Tracheophyllum.
- Petitrichia: These are instantly noticeable by their upturned, funnel- or bell-like bodies that are supported on a stalk. The opposite side of this stalk may be connected to particulate matter, like a sludge floc, which acts as an anchor for the organism. In other microorganisms, the stem is contractile, whereby the wider side of the bell function as an oral opening for the Peritrichia. Plus, they could have cilia spread around this structure to help in feeding. Some common peritrichs are Opercularia and Vorticella.
- Suctoria: This form of Ciliophoraonly develops cilia early in life, where these projections allow the new cell to disperse from its parents. Thereafter, they may lose their filaments and grow a stalk or feeding appendages. The stem is non-contractile and connects to particulate matter, while the feeding appendage is able to capture and feed, on nearby protozoa. This is done by infiltrating them and drawing in organic matter from the cytoplasm, so as to develop food vacuoles. Suctoria that can be found in wastewaters isAcineta and Podophyra.

Composition: Ciliates have a single or several macronuclei and micronuclei. The first manages metabolic and growth functions, whereas micronuclei are vital for reproduction.Multiplication is typically asexual, where cells divide into new ones through budding or binary fission. However, for a few ciliate species, genetic exchange between organisms may result from conjugation or autogamy which involves nuclear restructuring within an organism (Sukhareva-Buell, 2003).

Even though a majority of ciliates exist as free-living and marine species, like the Paramecium, others are ectocommensals, which live innocuously on the appendages or gills of invertebrates, whereas some like the dysentery-borne.

Ciliates are closely related to Apicomplexa and dinoflagellates. These are unicelled organisms, featuring an abundance of cilia or flagella on their surfaces. For their nucleus, the lesser micro-nucleus comprises of a diploid-germ nucleus, whereas the second bigger macronucleus has 1000s of replicas of just the biologically active genes. Ciliate nuclear gene arrangement can be really remarkable; genomes may not just be divided by introns or multiple short intervening chains, but the genetic fragment order might as well be scrambled. Hence, comprehensive editing is needed during production of the macronucleus, so as to generate the gene's active working replica. Even though the method by which this happens is still unclear. The cilia could be less in number, particularly in sessile species, or arranged into larger complex ciliary filaments, such as cirri. But the only big group which doesn't always have cilia on the body is Suctoria (Sukhareva-Buell, 2003).

They are sessile hunters whose dispersion stages are, nonetheless, ciliated. This unique group of cilia is easily identified by its feeding appendages. It's also good not to confuse tiny, flagella animals with ciliates as they may appear similar. The size scope of ciliates overlaps with various metazoan groups, like gastrotrichs, turbellarians and rotifers. Furthermore, some ciliates have mixotrophic properties, characterized by the existence of endosymbiotic algae, which acquire chloroplasts from consumed algae that remain practical within the ciliate cytoplasm.

In general, ciliates are divided into 12 groups. Those belonging to the class called Karyorelictea are considered the most primeval, comprising of multiple non-dividing macronuclei which aren't highly polyploid. These are mostly benthic, with a popular freshwater species being Loxodes.

6.2.2 Amoebae

Amoebas belong to the Sarcodina group of microorganisms. It's a large group comprising of members in almost every ecosystem imaginable, they usually develop by passing through a trophozoite phase that's bare, meaning the cell doesn't have any structural elements on its membrane which support the organism's shape. As such, what forms is an amorphous globule that travels through pseudopod projections. Amoebae are scattered globally plus are quite common in freshwater, soil and similar habitats. Although mostly harmless, some have been known to cause human-related illnesses (Wiser, 2010).

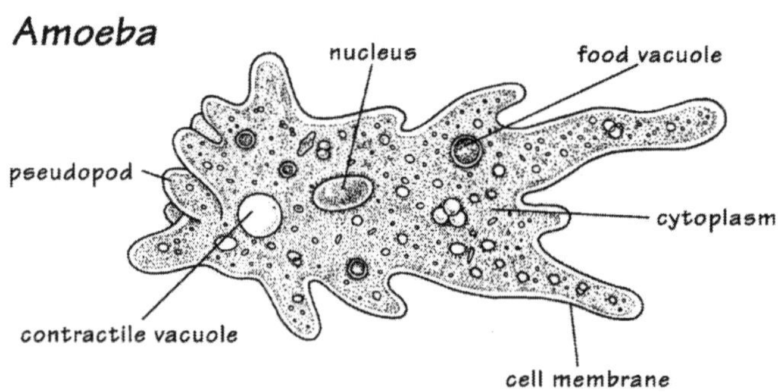

Figure 6.3: Shows the structures of an Amoebae.

Source: https://www.pinterest.com/pin/370843350540152881/

Distribution and behavior: Entamoeba histolytica, a form of amoeba, is mostly responsible for amoebic dysentery and liver abscess conditions. This is a universal health concern, responsible for more than 100,000 deaths each year. Likewise, the infection is quite predominant in kids compared to grownups, plus is commonly seen in tropical and sub-tropical parts of the world.On the other hand, Acanthamoeba is a more versatile amoeba thatsporadically infects people, where it may lead toamoebic keratitis or granulomatous amoebic encephalitis (Wiser, 2010).

The lifespan of both these amoebae specimens involves one host, plus is marked by an active consumption and division trophozoite phase, including a dormant cyst phase. The latter stage is usually found in polluted soil or aquatic systems, or even on dirty palms of food handlers that can last in the ecosystem for multiple weeks to months.

Pseudopodia: Most amoebae are capable of creating pseudopodia, which they use to move and forage for food. These temporary body protrusions may vary between organisms, depending on their role that relates to actin and myosin. These pseudopodia work in the amoebae similarly to their functions in the extension of vertebrate muscle. They are stimulated by ATP-based energy and various cations, such as magnesium and calcium. These molecular functions create a localized contractile reaction in the cytoplasm, whereby the cytoplasm turns inside out at the plasma film and travels posteriad into the cell, developing an outer shell of cytoplasm called the ectoplasm (Wiser, 2010). At the periphery of the cell, myosin and actin get disjointed; while the ectoplasm goes back to a relaxed form, becoming more liquid, and travels inward to develop endoplasm.

Once the endoplasm jets forward from the pressure coming from the contracting ectoplasm, actin once more is associated with myosin to produce afresh the contractile state. This entire effect, then, is a recurring outward and a posterior stream of ectoplasm off from the route of movement, including concomitant propulsion of the endoplasm away from the back of the cell towards the side of the developing pseudopod.

Physiologically, pseudopodia can be grouped into four main classes which are;

- Rhizopodia
- Axopodia
- Filopodia
- Lobopodia.

Nevertheless, the latter is the most common type among parasitic amoebae, they are blunt and might consist of endoplasm, ectoplasm or even both components together. For most species, lobopodia develops gradually. Studies on living specimens have shown a solid stream of molecular endoplasm, when existent, into the wider projection. However, Entamoeba histolytica, which is a common parasite present in the human gut, is unique in that lobopodiaisformed promptly and drawn back almost as quickly (Wiser, 2010).

Another exemption is seen among those amoebae which seem to travel on a substratum, but without any clear cytoplasmic protrusions. These amoebae called limax form borrowing from the slug, whose motion seems to resemble the development of pseudo-podia by trophozoites. Additionally, some flagellates can manage pseudopodial movement during certain stages in their life plus, equally, some amoebic organisms possess flagella throughout their developmental stages. The general rule, nonetheless, among flagellates is that the mode of movement is flagellar, whereas in amoebae it's pseudopodial.

A good majority of amoebae can't swim though; pseudopodial propulsion requires a substrate by which these microbes can slide. Parasitic species, thus, are normally found in alimentary ducts of their hosts, closely related tothe epithelial membrane (Wiser, 2010).

Appearance and behavior: The look and internal composition of pseudopods may be applied to differentiate between separate forms of amoebae. Amoebozoan types, like those in the Amoeba genre, usually have bulging pseudopods, circular at the peripherals and reasonably tubular in cross-section. Conversely, Cercozoan amoeboids, like Euglypha and Gromia comprise of slim, thread-like pseudopods. Foraminifera produces fine, multifaceted pseudopods that blend with each other to develop net-like structures. Some forms, like the Heliozoa and Radiolaria, have solid, needle-like structures, producing axopodia that's sustained from inside by packs of microtubules (Lwoff, 2014).

Moreover, to control osmotic pressure, freshwater amoebae usually have a flexible vacuole which ejects extra water from the organism. Marine amoebae, however, don't usually have a contractile vacuole since the numberof solutes within the organism isin equilibrium with the tonicity of their surrounding waters.

6.2.3 Flagellates

Flagellates are a group of Mastigophora that swim by waving their extended, whip-like flagella. This protozoal flagellum is physically different fromanother bacterial flagellum. Plus, the protozoal flagellum surfs while in motion while bacterial flagellum spins.There are various known illness-causing flagellates. For example, Trypanosoma brucei which causes sleeping sickness diseases, that's estimated to kill over 65,000 individuals in Africa each year. The lifespan of Trypanosoma is impressive since it uses a pair host. The infection manifests in people but must be transferred through a bite from a diseased tsetse fly (Lwoff, 2014).

Similarly, Giardia is a well-known pathogenic flagellate thatcauses diarrhea plus it's known casually as Beaver Fever. Yet another widespread protozoon is Trichomonas, which is a flagellate that causes urogenital syndromes in infected women.

Features: Flagellates are known for their possession of at least one or several flagella, these are long, narrow, hair-like attachments which function like organelles for movement and feeding. In free-ranging taxa, in comparison to other parasitic species, the number of flagella is rather restricted. Paramastix, for example, has a pair of 8–12 flagella, though most others comprise of 1–4. Naturally, where there are two flagella, one might project forwards, while the other trails from behind. Usually, the cell's flagella are lengthier than its body.

There are various classes of heterotrophic flagellates thatexist in freshwater, some common examples being kinetoplastids, choanoflagellates, bicoecids and diplomonads. Furthermore, some amoeboid forms, like cercomonads, schizopyrenida and amoeba-flagellates, consist of flagella but are regarded more of amoebae rather than belonging to this group.Other classes of flagellates may contain wholly or completely autotrophic forms such as chloroplasts. Nonetheless, most of the colored, the autotrophic genre isfurther capable of phagotrophy, which results inan overall condition known as mixotrophy, plus among these classes are some completely heterotrophic species (Lwoff, 2014).

The groups comprise heterotrophic and mixotrophic taxa; including chrysophytes, dinoflagellates, cryptophytes and euglenoids, plus they're usually regarded as phyla. Coloration and chloroplast form are crucial taxonomic characters within these groups.

Types of flagellates:

- Bicoecids: They resemble choanoflagellates in appearance, but lacking a collar. Moreover, just like choanoflagellates, they're covered in a lorica plus comprise of a flagellum that's applied for creating a feeding current. Plus, a second flagellum is found along with the cell, where it persists posteriorly to change into an attachment for the lorica's base.
- Kinetoplastids: These are known predominantly as parasites, particularly Trypanosoma and its families, though many members of its subgenre Bodina stay in freshwater. Among the best-known species is Bodo, which, same as bodonids, has a pair of flagella. One trailing, as the other projects ahead.As for cryptomonads, they include multiple common autotrophs and heterotrophs including a few mixotrophs. Their flagella pair are unequal in size and emerge from a sub-apical invagination normally called a "gullet." Even though it doesn't seem to be the spot of absorption

in heterotrophic strains. The pellicle is protected with plates, but these also aren't fully visible.

- Choanoflagellates: Commonly referred to as collared flagellates, these are distinct for their collar that encircles the single flagellum. These cells also have a close semblance to sponge choanocytes. Likewise, most choanoflagellates connect to the substrate or may be colonial, plus many have an exterior, loose-fitting casing or lorica, but this might be hard to see using a simple light microscope.
- Dinoflagellates: They form a large and distinct group, that's perhaps more pronounced in marine compared to fresh-water environments. Their distinctive arrangement of flagella, whereby one spirals around the organism in a girdle, whereas the second is distally directed around another groove knowns as sulcus, makes dinoflagellates quite impressive looking. Yet again, mixotrophy and heterotrophy are typical. Casing of plates may be present or absent, thus the phrases armored and bare dinoflagellates.

Form and behavior:Flagellates are typically supported by microtubules in a distinctive arrangement, where 9 fused pairs encircle 2 central singlets. They emerge from a basal body. Moreover, in some flagellates, the flagella move food into the mouth or cytostome, and it's ingested from there. Flagella often contain hairs, known as mastigonemes, or may have rods. Plus, their ultrastructure plays a key role in categorizing eukaryotes (Lwoff, 2014).Among protoctists and microbial animals, the flagellate is a creature with one or several flagella. A few cells found in some animals could be flagellate, such as the spermatozoa present in most vertebrates' phyla. Flowering plants, however, don't generate flagellate cells, though green algae, ferns, mosses and a couple of gymnosperms which closely resemble plants do so. Equally, most fungi don't create cells with flagella, though the primeval fungal chytrids does.

6.3 ROLES OF SOIL PROTOZOA

6.3.1 Mineralizing Nutrients in the Soil

Protozoa play a significant role in the mineralization of nutrients, by making them ready for use by plants or any other soil organisms. Generally, Protozoa have lower amounts of nitrogen within their cells when compared to the bacteria they consume. In fact, the quotient of carbon-to-nitrogen within protozoa is 10:1 (Corliss, 2016).

Conversely, bacteria consumed by protozoa contain excessive nitrogen for the quantity of carbon the protozoa require. They churn out this excess nitrogen in form of ammonium, a process that typically occurs around the plant's root system. While bacteria and other soil germs may rapidly consume this ammonium, some of it is actually recycled by the plant.

Figure 6.4: Protozoa can enrich the soil and make it more fertile for plants to grow.

Source:https://plantlet.org/soil-microbiology-the-scientific-discipline-of-microbial-communities-in-soil/

In scientific studies, when protozoa were introduced into the soil, nutrient amounts consumed up by crops improved by 75%. Sugar additions further caused an 18% surge of overall nutrients in the shoots in case the protozoa were present, though this had little effect in the absence of bacteria. Moreover, plants with considerable root weight are better placed to uptake inorganic nutrients when protozoa are present. These tests show that, with energy delivered by plant-roots or even external glucose supplementation, protozoa can mineralize nutrients from the soil's organic matter to support growth processes. The function of protozoa in cycling of nutrients and energy flow, however, is largely based upon their bionomics. Some distinct features of protozoa, like their high reproduction rate, also play a part in mineralization (Corliss, 2016).

Nutrient mineralization in the soil helps to create more essential minerals which the plant needs to germinate in a healthy way. Without these components, the crop may easily wither away due to a lack of proper nutrition.

6.3.2 Regulating Bacteria and Algae Populations in the Soil

Protozoa are single-celled organisms that feed predominantly on bacteria present in the soil, though they can also consume other protozoa, fungi or sometimes even soluble organic matter like algae. In terms of size, protozoa are multiple times bigger than bacteria, usually varying in size from 1/5000 up to 1/50 an inch in width. As they consume bacteria present in the soil, protozoa emit excess nitrogen which may then be utilized by crops or other creatures existing in the food web (Corliss, 2016).

Some protozoa are useful in nutrient cycling and promoting photosynthesis as well as plant growth. Their role, however, is determined not just by their size or weight but also by their catalytic reaction within different ecosystems. These organisms also have an influence on various nutrient cycles, such as the carbon cycle, sulfur cycle, phosphorus cycle and nitrogen cycle where organic energy are implicit.

The geometry of nutrient cycling is determined by the nature and distribution of sources and sinks, the most important being the atmosphere, the ocean and soil, and the availability of the major, minor, and trace elements, and other growth factors. Nevertheless, the axis of nutrient cycling may also be driven by nature, including the distribution of nutrient sources, some popular ones being atmospheric conditions, the soil and availability of primary, secondary, and trace compounds, including other growth factors.

Likewise, protozoa play a crucial role in managing bacteria populations. When they consume bacteria, protozoa help to balance bacterial numbers, which in turn enhances the decomposition capacity and soil aggregation. Basically, protozoa require bacteria for food and water for habitat in order to survive (Corliss, 2016).

6.3.3 Act as an important Food Source for other Soil Organisms

Protozoa are a vital food source for other protozoa and microorganisms present in the soil. They are usually found at the topmost layer of soil, around 15-20 cm from the surface. Plus, being bigger than most bacteria or fungi in size, they typically inhabit the places around soil pores instead of the tiny pores occupied by bacteria. A considerable number of protozoa are present in freshwater, where they travel freely. In the soil, their locomotion is considerably restricted, plus they are generally smaller in dimensionswhen compared to other freshwater microbes. Fungal spores, in particular, feed on protozoa and play a significant role in nutrient synthesis. Within the soil, protozoa numbers usually escalate around nutrient hotspots like the rhizosphere, or in places where nutrients have already been availed. Some soil microflora, like soil nematodes, may also feed on them (Corliss, 2016).

Protozoa generally flourish in humid, well-drained soils plus are most frequent in surface horizons, with their biomass varying from 20 up to 200 kg/ha. The larger groups of protozoa may be identified by their morphology, including how they feed. However, amoebae are perhaps the most abundant of the protozoa. They ingest food and travel via temporary extensions found on their pseudopodia. Furthermore, some of them, like testate amoeba, emit or use matter from the ecosystem to develop a tough shell that safeguards them against predation or other environmental hazards.

When soil microorganisms consume amoeba and other protozoans, they usually release nitrogen as a byproduct of their metabolism. This compound is useful in developing proteins and nucleic acids, which are essential in development (Corliss, 2016).

6.3.4 Help to Suppress Disease by competing with or Feeding on Pathogens

Among the major ecological significance of protozoa is their role in pathogen suppression. They help to control bacterial populations, which may otherwise be dangerous to plant populations. This is crucial in maintaining biological balance and diversity. Similarly, both aquatic and land-based protozoa accelerate the decomposition of plant and animal remains in the ecosystem so that they don't act as breeding grounds for diseases, which may spread into plants and cause illnesses. Furthermore, decomposition is an essential ecological process since vital nutrients and minerals get reverted back into the ecosystem. Protozoa consume organic matter produced by decaying plants or faunas, thereby facilitating the breakdown of molecules into simpler nutrients that plant can uptake (Corliss, 2016). Also, predation by protozoa has been found to promote the generation of plant growth hormones and improve the survival chances of helpful microbes that suppress pathogens. Protozoa further stimulate the creation of compounds associated with disease suppression, like antibiotics and siderophores. Consequently, it's likely that co-inoculating helpful bacteria together with bacterivorous protozoa could result in enhanced plant growth and vitality.

Some species of protozoa exist in mutual relationships alongside certain plants. For instance, photosynthetic flagellate protozoa may supply crops with a regular dose of carbohydrates, which are created during photosynthesis. A high percentage of the plant's energy requirement may be provided by these protozoa. In equal measure, the crop gives the microbes a place to call home and constant protection. Protozoa may further act as bioindicators in the soil, by monitoring its health and integrity to ensure ecological integrity. Microbes sampled from the soil can show if there are diseases or toxins that can potentially damage plants.

6.4 NEMATODES IN THE SOIL

Nematodes, also called roundworms, are found in the Nematoda phylum. This is a comprehensive animal phylum covering a wide array of habitats. In terms of taxa, they are categorized alongside insects including other molting creatures found in the group Ecdysozoa. But unlike flatworms, they have tubular gastrointestinal systems with orifices at both ends.

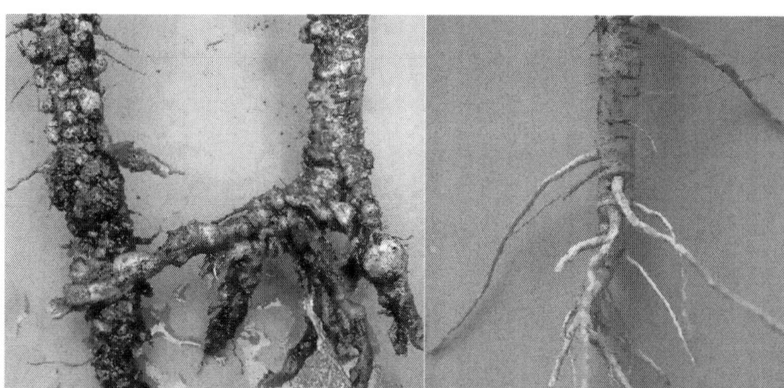

Figure 6.5: A bulbous and sickly-looking root-infested by root-knot nematodes, in comparison to a healthier root on the right.

Source: https://www.infonet-biovision.org/PlantHealth/Pests/Root-knot-nematodes

Nematode species are quite numerous in the environment. In fact, a 2013 survey indicates that there are around 25,000 types in existence today, while others place the amount at around 40,000 species globally. These creatures have effectively adapted to almost every environment, be it marine or soil, the highest or lowest elevations. Studies show that they represent approximately 90% of all creatures on the seafloor, plus for every human being, there are approximately 60 billion nematodes in existence (Lwoff, 2014).

6.4.1 Activity is Soils

Most soil nematodes are found in the rhizosphere section where their abundance typically declines with increased depth and farness from the plants. As researchers have found, around 70 percent of the microbial- and fungal-eating nematodes exist in the rhizosphere, specifically the portion of soil that's 1-2 mm just off the root surface or rhizoplane. In a 1993 study, it was discovered that nematodes traveled to packets of decaying grass deposits, where there were significant quantities of labile substrates including microbial food sources. The scientists concluded that they search out for these "hot points" of compact organic matter, unlike protozoa which don't exhibit similar behavior. Moreover, the distance and specific position that a nematode occupies from plant roots will depend on the specific species, soil type, moisture levels and temperature (Lwoff, 2014).

6.4.2 Five Broad Groups Nematodes in the Soil

Bacterial-feeders' nematodes: Nematodes have evolved various behaviors that primarily serve the role of locating food in the ecosystem. Different nematodes exhibit their own unique dietary choices, some hunt for particular types of bacteria while leaving those that are a bit hard to find. The food-seeking activity is more pronounced in nematodes that have already established a dietary trend. When hunting for bacteria, they may alternate between two methods of locomotion, commonly referred to as dwelling where movement is with constant breaks and reversals, as well as roaming which is characterized by straight rapid movement.

Almost all known Bacterial-feeders' nematodes behaviors are influenced by food. For instance, in Caenorhabditis elegans which is a transparent free-living nematode measuring about 1mm in length, it's been established that bacterial food is the main sensory stimulus. The function of bacterial food as reinforcement for certain nematode behaviors like foraging has well been documented. C. elegans feed almost entirely on bacteria and eat almost constantly throughout their life cycle, apart from the molt and diapause phases. To accomplish this, they identify food in their ecosystem, travel towards it, and after coming close launch pharyngeal pumping mechanism which allows them to consume the bacteria. Though at first glance this may look like a basic repertoire of behaviors, in fact,C. elegans nematode just like other complex creatures makes complex decisions regarding food choices. For instance, their feeding pattern is predetermined by their nutritional composition, plus they might learn how to avoid pathogenic microorganisms in the environment. Interestingly also, male bacterial feeding nematodes, have been captured abandoning their food to look a female, which is behavior commonly seen in larger animals (Sukhareva-Buell, 2003).

Fungal-feeder nematodes: Fungal-feeding nematodes serve as biological management agents of soil-based fungi, which particularly helps to diminish the contamination risk on sprouting plants. Case in point, both the genres Aphelenchoides and Aphelenchus have been found to inhibit damping-off triggered by R. solani fungi commonly found in cauliflower seedlings. Plus, they help to boost infection-suppressive qualities in compost.

Most fungal-feeders of this species belong to the Tylenchidae family, which is an expansive class of soil nematodes whose feeding trends are still being studied. In studies, measuring fungal-feeding behaviors of nematodes found within the genus Filenchus. It was discovered that Filenchusdiscrepans and Filenchusmisellus nematodes had moderate to massive population growth rates, when fed on saprophytic fungi but reduced population growths when given plant-pathogenic fungi like Pythium ultimum and Fusarium oxysporum. This suggests that the nematodes could be having preferences on the kind of fungi they prefer consuming (Sukhareva-Buell, 2003).

Fungivorous nematodes are valuable organisms for fast-tracking matter decomposition, plus they serve as bio-control agents of crop pathogenic fungi as well. The A. avenae nematode in particular, which thrives in temperate climates, has been applied for the bio-management of soil-borne crop diseases infected by fungus. Moreover, industrial wastes may be used efficiently in mass production of A. avenae, plus repeated application of substrates which could still be having nematodes can further limit population numbers. This is inclusive ofother fungal pathogens like Pythium, Phytophthora, Fusarium, Rhizoctonia and Verticillium. In soil, the effects of fungal-feeders were quite significant in controlling proliferation, though not steady, across different nematode species. Some researchers further hypothesize that plant-feeding nematodes could have evolved from ones feeding on fungi (Sukhareva-Buell, 2003).

Predatory nematodes: Predatory nematodes are tiny microscopic creatures that feed on other organisms, or may even revert to cannibalism. They are wormlike in shape, plus the adults usually are referred to as "double-death" predators. These microbes control the population of a wide range of bugs, many of which have life phases in the soil such as beetle larvae. Some of the creatures they feed on include; cutworms, gypsy moths and cucumber beetle among others. Some types of predatory nematodes have hermaphroditic features, whereby both male and female reproductive organs are found within the same creature. As for their eggs, they are among the toughest known to science with the ability to withstand freezing temperatures, or even go without oxygen for long without getting spoiled (Sukhareva-Buell, 2003).

Predatory nematodes mostly attack soft-bodied, and slow-traveling insect pests, but they don't affect plants or earthworms. These also are the only physiological control mechanisms that work beneath the soil surface. As a farmer or gardener, simply apply them in a spray container or fertilizer injection structure when soil-temperatures are over 55° F. A single package of approximately 1 million predatory nematodes is enough to treat around 3000 square feet of land where they will feed on disease-causing microorganisms, which would have otherwise devastated your crops. Microscopic attacking nematodes have the ability to eradicate almost any bug that stays in the soil, including those as big as cutworms. As long as the target insect pest spends a considerable amount of its lifespan resting in the soil, the nematodes will attack by first invading the bug's body and then reproducing more nematodes on the insects' carcass.

Omnivore nematodes: As their name suggests, omnivore nematodes feed on both plant matter and microbes existing in the soil. Food habits of nematodes present in soil are comparatively specific. For instance, bacterivores may eat only bacteria and plant roots, whereas other omnivorous nematodes would feed on protozoa instead of bacteria and plant matter.

In their natural ecosystem, the soil free-ranging omnivore nematode community structure, size, and activity levels are determined by abiotic factors, including moisture, temperature, organic matter portion and pH levels. At deeper soils, the nematode concentration often diminishes, as observed in studies of soil layers between 0-30 cm in depth. This remarkable difference in populations is predictable since the disparities in nematode populations at lower points are less marked, plus they can happen as a result of natural edaphic features, root development and dissemination, as well as anthropogenic disturbances. Environmental and climatic conditions further play a role in determining the behaviors of omnivore nematodes, when these factors are favorable, they may engage in more feeding than when they aren't conducive.

For instance, in humid Mediterranean location, omnivore nematode populations reached their peak during winter, whereas in arid and semi-arid areas a nearly opposite phenomena were observed. Whereby the lowest abundances occurred during winter season, presumably as a result of nutrient washout coming from the soil-profile emanating in the rainy season. Furthermore, free-ranging nematodes are usually affected by biotic and abiotic factors in the environment, such as mineral levels. Sodium and Potassium concentrations in particular affect pH values, which in turn determine the nematodes' feeding habits and availability of plants and microbes for consumption (Sukhareva-Buell, 2003).

Root-feeder nematodes: Root-feeder nematodes got discovered first in 1855. The scientist responsible for identifying them, Berkeley, noted that they were causing destruction on cucumbers. However, it wasn't until 1949 that another scientist named Chitwood identified 4 different species of nematodes, including 1 subspecies in the genre Meloidogyne.Like most plant-parasitic nematodes, root-feeders have a special stylet for inputting secretions, including absorption of minerals from host crop cells. Nematodes don't have any internal skeletal structure, plus their "skin" or outer cover protects against internal turgor weight to help preserve body frame and aid locomotion.

Likewise, different from other plant-parasite nematodes, their females are generally globose and inactive at maturity. They vary in diameter from 400-1000 μm. Upon establishing a feeding site, these microbes permanently stay at that specific spot within the root. Their feeding location is really a cluster of cells called "giant-cells". Once a nematode penetrates through a plant cell using the stylet, it inserts secretory proteins thattrigger changes inside the parasitized cells. These rapidly become multi-nucleate, meaning they contain several nuclei, as nuclear multiplication happens in the nonexistence of cell-wall formation. This procedure is regarded to be "un-coupled" from cell multiplication (Sukhareva-Buell, 2003).

Root-feeder nematodes typically start their lives in egg form but rapidly grow into J1 or stage one juvenile nematodes. This J1 phase resides especially within the translucent egg casing, where it grows into the next level of J2 nematode. They attack maturing root tips and penetrate roots inter-cellularly, meaning from behind the root-cap. Consequently, the materials move to the site of cell-extension, where they launch a feeding spot by introducing esophageal gland emissions into host root cells.

6.4.3 Role of Nematodes is soil

Nematodes play a crucial role in the ecosystem since they help to decompose organic materials, as well as recycle various essential nutrients found in the soil. These organismsfurther affect the decomposition and mineralization of biodegradable matter. Not to mention, they often form relationships with bacteria and fungi living in the soil, which usually help in performing the basic roles of the nematodes. Likewise, nematodes occupy a tiny fraction of the entire biomass and respiration present in the soil. Meaning, they are also crucial for the generation of carbon dioxide within the soil. The carbon dioxide further helps bacteria and fungi in decomposing the organic material in the soil (Lwoff, 2014).

Biotic indicators: Nematodes have some unique biological features that make them useful as indicators. For instance, they consist of a permeable cuticle, allowing them to react with a set of responses to pollutants, while further corresponding with the restorative quality of soil ecosystems. Secondly, some nematodes comprise of resistant stages like cryptobiosis or cysts, which allow them to live inactively in ecological conditions otherwise unfavorable for growth and expansion. Also, there are certain nematode taxa like Dorylaimidae with no resistant stages, but maybe quite sensitive to ecological changes. Furthermore, nematodes have special heat-shock proteins which are greatly conserved. The manifestation of these proteins is improved when set up to factors like heat, biological toxins and metal ions. Conceivably, these proteins may serve as biomarkers for the ecotoxicological evaluation of soils.

Up from the 70s, nematodes have been applied as ecological bio-monitors for marine systems as well. Case in point, Panagrellus redivivus is a typical strain that's been used for biomonitoring and detection of toxin concentrations affecting molting. Likewise, organism size via stimulation, lethality or inhibition provides a quick bioassay thatusually costs less than ten percent of the Salmonella bioassay. Currently, this nematode has determined the toxic effects of around 400 single chemicals present in the soil (Lwoff, 2014).

Nutrient recycling and mineralization:Nematode communities are recognized for playing crucial roles in ecosystem processes, like nutrient cycling, while also responding to the ecological distribution of biomass found in plants as well as nitrogen mineralization. In laboratory experiments, extra nitrogen is present in form of ammonium when bacterivorous or fungivorous nematodes exist in the soil sample than when missing. Nitrogen mineralized from microbial grazing is consequently available to plants, plus it's been shown to affect biomass distribution in plants. For microcosm tests with buffalo grass, it was established that plant shoots became larger in soils where fungi, bacteria, and other respective grazers were present than in those with less defined soil food-webs. Root biomass can further increase in the availability of microbial-grazing nematodes.In field conditions, predatory nematodes and bacterivorous have been estimated to cover about 8-19% of all nitrogen mineralization processes, both in conservative and integrated farming methods, correspondingly.

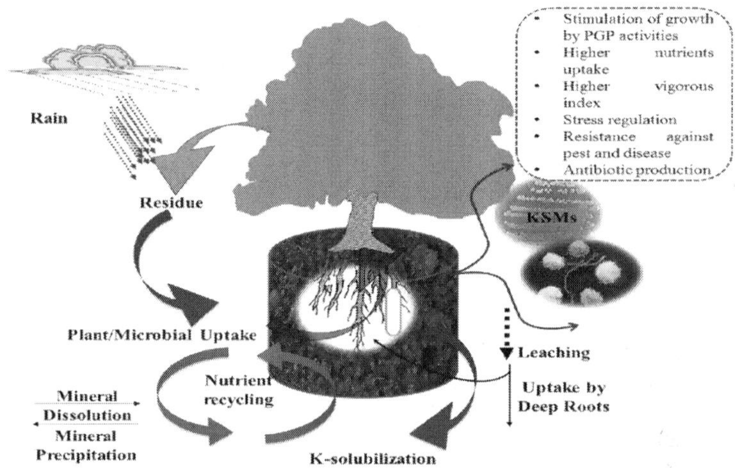

Figure 6.6: Nematodes play a role in nitrogen mineralization indirectly through feeding on decomposer microbes, emitting ammonium, and controlling nitrogen levels in live biomass.

Source: https://www.researchgate.net/figure/Plant-growth-promoting-PGP-activities-of-potassium-solubilizing-microbes-KSMs-for_fig2_328430236

In particular, predatory nematodes further regulate nitrogen mineralization through consuming microbial-grazing nematodes, which serves as a conduit for passing resources from the bottom to apex trophic levels (Lwoff, 2014).

6.4.4 Managing Parasitic Nematodes

Most crops are affected by plant-parasitic nematodes [PPN], given their abundance in the environment. In fact, these creatures are more intense and complex in warmer areas compared to cooler ones, including horticultural, mono-culture and plantation crop settings. During the olden days, nematodes caused many farmers to migrate from their farms as a result of soil sickness.

Identification: These nematodes are quite widespread in cultivated crops, which are typically attacked by at least one or more strains. The organisms are usually microscopic normally measuring around 0.5-1mm in length. Plus, they are tubular in shape, with the body narrowing towards their head and bottom ends. Likewise, mature females of various predatory nematodes may have swollen pouches, resembling a lemon-pear kidney or uneven shapes whereas males remain slender. Nevertheless, the most common diagnostic attribute of PPN is they comprise of a spear or stylet organ at the mouth-end. This is an elongated, sharp and pointed tubular structure resembling a physician's hypodermic needle. It's typically moved back and forth inside the plant tissue. The nematode draws up plant sap while at the same instance injecting a tiny amount of its gland emission, comprising of digestive catalysts into the tissue. Generally, the plant's expression of different disease symptoms can be attributed to toxic components injected, as well as the reduction of plant nutrients (Lwoff, 2014).

Figure 6.7: Sampling for parasitic nematodes.

Source: https://www.plantmanagementnetwork.org/elements/view.aspx?ID=5419

Management techniques: All parasitic nematode control strategies are directed towards cutting down their population in the soil, which should be low enough for crops to grow proficiently. Generally, a blend of control techniques is required to grow crops successfully. Nematodes are omnipresent, with the ability to be transmitted via various methods. Though irrigation-water is the leading source of their transfer, followed by the usage of infested composts and soil.

- Crop rotation: It's among the most crucial methods used for keeping nematode populace under check, particularly when there are distinctive host preferences such as; cyst nematodes found in wheat and potatoes. Vulnerable crops must be grown just a few times in the year, regularly revolving them with non-host plants. Vegetables shouldn't be grown recurrently in the same space but should be swapped with cereals so as to minimize invasion with root-based nematodes.
- Fallowing and plowing in summer: Hot and dry environments are ideal for managing nematode populations during summer period. These creatures are normally sensitive to hot weather and dehydrating effect of the sun as well as wind. Therefore, ensuring that the land remains fallow in summer season, including deep-plowing of the soil around 2 to 3 times in intervals for 10-15 days, maybe an excellent way to minimize soil nematode populations and other organisms.
- Heat treatment: It's a common procedure for soil-sterilization, which serves to protect against all harmful microorganisms. Plant nurseries use autoclaved soil for potting, where the soil is treated with heated water at pressures of around 30 psi for half an hour. For greenhouses, the vapor is released via perforated pipes spread throughout the soil. Temperatures could rise up to 82°C with the whole process lasting for around 30 minutes. Equally, soil dry-heating through burning off the residue crop waste can be useful in controlling nematode populations.

Chapter 7

Soil Health Management Practices and How to Sustain Beneficial Soil-Borne Pathogens

CONTENTS

7.1 INTRODUCTION

Soil health is asoil condition relating to its ability in maintaining environmental conditions that can sustain humans, plants and animals. The definition brings out the importance of soil management. In soil there are living organisms that need to be provided with the basic necessities of living that includes-shelter food and water to perform their function of producing food. Soil is a living ecosystem since only living things can have health hence it should be managed by taking care of it. Soil should be managed in order to give nutrients for the growth of plants. Soil health is linked to environmental sustainability and land production. Soil is said to be healthy when it brings in profit and has good production.

The following are some of the major functions of healthy soil:

- To sustain life and society- Helps in the production of living things.
- Store and filter water- Microbes and minerals in soil play the roles in filtering, detoxifying and degrading inorganic and organic materials and industrial waste or by-products.
- Cycle and store matter- Nitrogen, phosphorus, and carbon are stored, cycled and are transformed in the soil.
- Providing physical support- A medium is provided by soil structure for plant roots. Soil also provides support to things constructed by humans such as buildings.
- Resisting erosion-healthy soil helps in controlling where water flows during irrigation and when it is raining or during snowmelt. Water flows into or through the soil.

Soil health brings in the idea that soil needs to be managed with care in order to maintain its capability to function properly. Soil is full of life. Soil can be managed by applying compost organic materials to it in order to reduce plant diseases and pesticides, by using covercrops to protect the health of the soil and also practicing crop rotation increases the crop yields.

Figure 7.1: Dynamic soil quality: beneficial vs unfavorable managing.

Source: http://www.oneorganic.com.au/blog/412-2/

Due to different management, soil health has taken important directions. The net farmer's profit decreases or there is loss when the soil does not function well. There is also decrease in production and decrease in environmental quality for a long period. This may be caused by limitations to soil processes occurring (Patel, 2016).

The importance of improving and maintaining soil health includes:

- The cost of tillage is reduced since its long-term
- Cost of inputs reduces due to proper use of pesticides, herbicides, fertilizer and proper irrigation
- The risk of loss is highly reduced in case of heavy rain, disease or pest outbreak.
- Plants grow healthy with better quality and yield.
- In the landscape, healthy soil acts as an environmental buffer.
- There are large pores of spaces that provide shelter to nutrients, organisms and water.
- Healthy soil does not acidify.

7.2 SOIL PROPERTIES

A soil has water, air, organic matter and mineral particles. When these are combined, it helps in determining the soil properties. These are three main types of soil properties; Soil physical property, Soil chemical properties and soil biological properties.

7.2.1 Soil Physical Properties

Soil physical property defines the movement of water, air and dissolved chemicals through the soil, also the conditions affecting root growth soil erosion processes and germination. The physical properties of soil include soil texture, soil color, soil structure, soil density, temperature in soil, porosity in soil and air (Patel, 2016). Soil color shows important properties, for example, organic matter, redox condition and water.

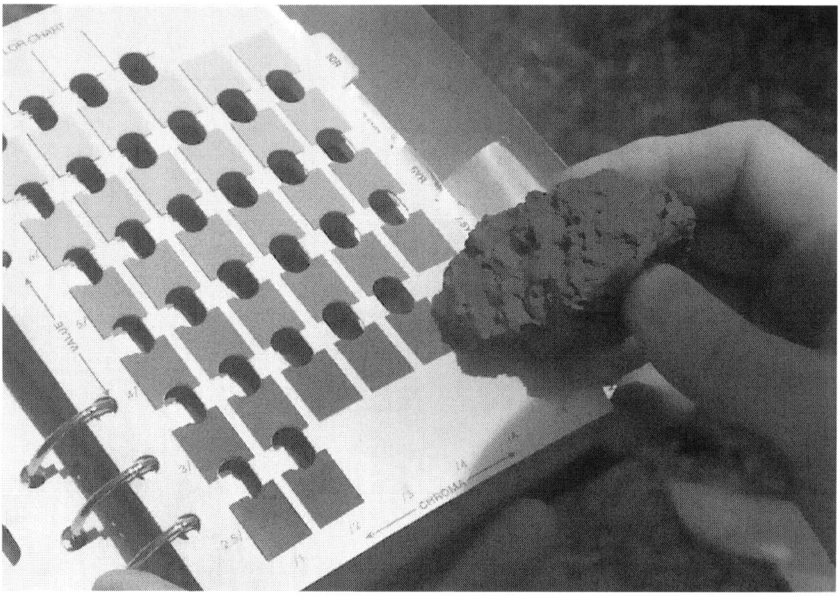

Figure 7.2: Measuring the value and color of the soil.

Source: https://www.soils4teachers.org/physical-properties

Primarily soil color is affected by soil minerals which show what is in a particular soil. For example, iron isa yellowish color or deep orange-brown color. Showing us these soil that has high organic matter is black or dark brown. Color also tells us the soil behavior. A brightly colored soil has good drainage while soil with yellow, red or grey soil patterns appears soggy and wet (Patel, 2016).

Soil particle alignment and kinds of soil particles are related in soil texture, soil structure, soil porosity, soil density and soil consistency. The two different types of soil particles include primary particles and secondary particles.: silt, sand and clay are primary particles that are grouped according to their diameter.

Soil texture is not easily altered. It is the relative proportion ofsoil. There are several textural classes that range from sand to soil. It is easy to assess texture of soil by mixing water and soil and making balls using your hand. You can use the below table to assess your soil.

Table 7.1: Properties of different soils

Texture class	**Properties of soil ball**
Sand class	It doesn't form a ball. It is not sticky and feels very sandy.
Sandy loam class	It forms a ball, ribbon of approximately 14-25 millimetreslong.it feels slightly sticky and sandy.
Loam class	It forms a ball with a regular and even surface with a ribbon of 25 to 40 mm. It is averagely sticky and feels slightly sandy.
Clay class	Forms a smooth plastic ball with a ribbon of greater than 75mm.It has a clay content of 40% and its very sticky
Light clay class	It forms an even and regular plastic ball of ribbon ranging from 51-75mm. It is very sticky.

Source: http://fertsmart.dairyingfortomorrow.com.au/dairy-soils-and-fertiliser-manual/chapter-4-soil-properties/4-2-physical-properties/

Soil particle alignment into separate geometric patterns is called soil structure. Soil structure is grouped into different types on the basis of shapes, grades instability including classes in size. The structure of soil is docile.

The development of soil structure is influenced by:

- The type and amount of primal matter, because it gives food for bacteria and fungi in soil.
- The type and amount of clay soil, as well as the ions exchangeable on the clay and acts with water as the medium between clay particles.
- Presence cementing agents of iron
- Binding between inorganic compounds and organic compounds that includes: aluminum oxides, clays

Soil structure and soil texture regulate density, retention, compactness, porosity and moving of water and air in soil together. The temperature in air is a little bit lower than the temperature in soil of an area. Soil temperature affects the soil biota living processes inclusion of plants. Soil pores keep possession of air in soil; its makeup is not consistent; it carries high moisture intensity and carbon dioxide plus oxygen concentration is less than air atmosphere. Soil air plays a major function in microorganisms and transformation of mineral, organic matter and in respiration of plant roots.

7.2.2 Soil Chemical Properties

The parent material determines the soil chemical content to a large extent which is produced conclusively. Leaching takes place in the lower part of the soil layer while other chemicals accumulate in the soil's upper layer. The chemicals which are swiftly removed include; sodium, sulfates, chlorides and calcium, magnesium and potassium follow them. There is very slow decomposition of the oxides and silicates of aluminum. The chemical properties of soil are very important. It is determined by its percentage humus, organic matter content, and iron which are rarely leached. When there is contact of air in the soil with some of these products, chemical reaction will occur, including oxidation which results in chemicals formation. These chemicals either dissolve or are highly damaged than the initial ones. This leads to increased speed of weathering process, more changes in the chemical composition and increased chemical leaching (Idowu, 2014).

There may be a development of a fresh sulfate acidic soil. When there is water logging, soil carrying iron sulfurs is exposed to air, this can occur during pond construction. There may be too much acid in the pond if the soil acidifies during the oxidation of iron-sulfur. Carbon dioxide is present in the air that is contained in soil. When water is combined with carbon dioxide it forms weak carbonic acid that later reacts with soil chemicals.

The release of nutrients to the plants is influenced by the extent of minerals in the soil. If there is a right balance, adequate nutrients will be produced by soils (Idowu, 2014).

Balanced soil chemistry: This begins with the balancing of minerals in the soil. The role of fertilizers is feeding the crop and to balance the chemistry of the soil. The best clay-humus content is about 20% magnesium, 70% calcium, and about 5% potassium that leaves 10% consistently that is able to be filled by hydrogen. The pH of the soil is determined by hydrogen and it is of importance to creating availability of the nutrients and keeping soil life active.

Figure 7.3: The chemical soil fertility is mostly resolved by the macro-elements' presence, the micro-elements presence or the trace elements, the soil acidity or pH, the salt content and the cation-exchange capacity.

Source: https://www.gartengnom.net/naehrstoffe-fuer-pflanzen/

The most vital plant's nutrients are macronutrients. They take place in soil naturally to some degree and manure; fertilizers and compost are added to supplement it. Calcium (Ca), Sulfur, magnesium (Mg), phosphate, potassium (K), nitrogen, sodium are the examples of the macro-elements. In micronutrients, nutrients that plants need in small amounts are trace elements. These elements include cobalt, silicon, zinc, boron, copper, manganese, iron and molybdenum (Idowu, 2014). The measure of the hydrogen ions(H^+) concentration without charge is known as Acidity (pH). A low pH is signified by a large soil concentration, while a high pH signifies little concentration. Soils aresaid to be acidic when it has a pH below 6, while base or alkaline are those with a pH above 6.

7.2.3 Soil Biological Properties

Nitrogen cycle: There is a relation in microbial and faunal activities to soil properties. The organisms are nematodes, protozoa, bacteria and different arthropods. An important role is played by soil biology in determining most of the soil behaviors. Little is known about the biology of soil and howthe nature of soil is influenced. Organic matter is broken down by the soil-borne pathogens and makes nutrients ready for uptake by plants by doing so (Magdalena et al, 2018). Leaching is prevented by nutrients stored in the body to avoid nutrients stored in the body of soil organisms from being lost. During bioturbation by the earthworms in the soil, soil structure is maintained by the microbes.

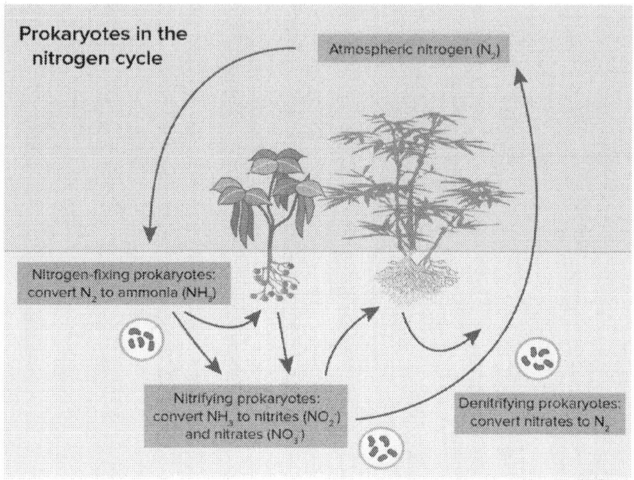

Figure 7.4: Nitrogen cycle.

Source: https://www.khanacademy.org/science/biology/ecology/biogeochemical-cycles/a/the-nitrogen-cycle

Bacteria play a crucial role in the nitrogen cycle hence affecting:

- Fixation of nitrogen which is carried out by rhizobia or by nitrogen-fixing bacteria in the soil or water such as Azotobacter. Colonies are formed in nodules by these bacteria.
- Nitrification in which nitrogen can be transformed in ammonium, which results from the proteins decomposition to nitrates, which are available for growth of plants.
- Denitrification returns nitrogen to the atmosphere. Anaerobes, Achromobacter and Pseudomonas inclusion tend to be denitrifying bacteria. The Nitrates and nitrites are converted in the soil to nitrogen gas or to compounds that are gaseous such as nitric oxide by purification process which is caused by oxygen-free conditions. When denitrification is immense, can lead to subsequent loss and overall losses of soil fertility and available soil nitrogen

- Mineralization is defined as the process in which decomposers and bacteria convert pure nitrogen to ammonium. When an animal expels waste or a plant or animal dies, the original form of nitrogen is organic. Organic nitrogen found in the remains is converted back to ammonium by fungi or bacteria in some cases by a process known as ammonification or mineralization.

Carbon cycle: This isa process by which the carbon element is exchanged between the geosphere, hydrosphere, biosphere and atmosphere of Earth. It is the most vital development on this planet since it allows reusing and recycling of its richest element. Carbon exchanges between reservoirs, the annual movements of carbon, happen due to a variety of chemical, biological processes, physical processes in the soil (Patel, 2016).The microbes that are in soil recycle nutrients such as nitrogen and carbon into the system of the soil. Microbes consume almost all of the organic material within the root zone or added to the litter. As a result, carbon has a reservoir with a quick turnover duration of 1 year to 3 years in various cases. These microbial consumption by-products arecarbon dioxide, water, and some of other compounds called humus.

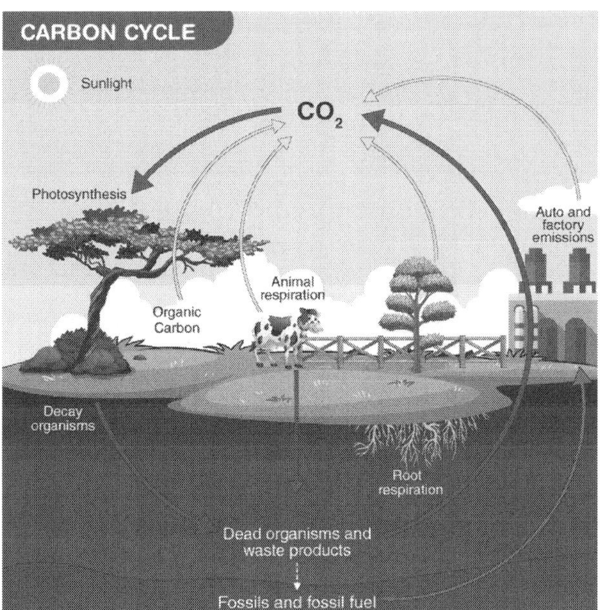

Figure 7.5: Carbon cycle in brief.

Source: https://byjus.com/biology/carbon-cycle/

7.3 SOIL FOOD WEB

Soil food web represents a special group of organisms living in the soil. In other words, a soil food web represents a system of living organisms in the soil and how they relate and interact with each other.

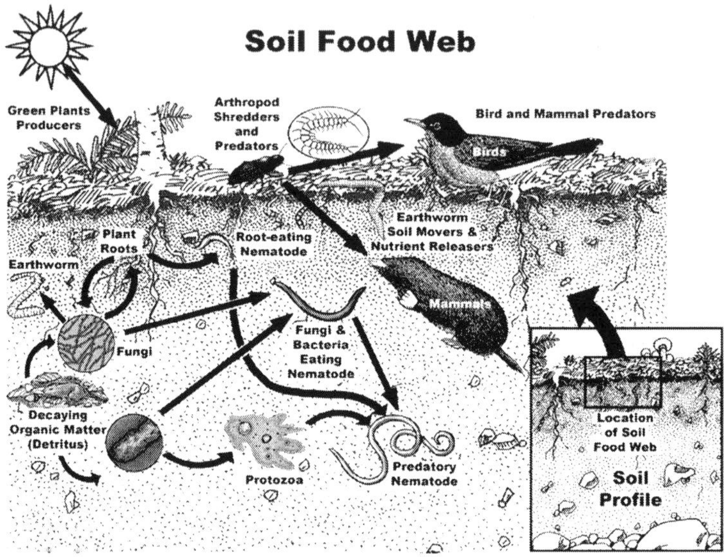

Figure 7.6: A diagram showing different conversions shown by arrows of nutrients and energy.

Source: https://www.exploringnature.org/db/view/Soil-Food-Web

The moss, lichens, plants, photosynthetic bacteria, and algae are primary producers that use energy from the sun in fixing carbon dioxide from the atmosphere fuel all food webs. Some bacteria chemoautotrophs are given energy by sulfur, nitrogen, or iron compounds. By consuming organic compounds that are found in plantsand waste products, most of the organisms of soil receive carbon and energy (Patel, 2016).

7.3.1 Importance of Soil Food Web

A healthy soil food web gives great ecological importance to all landscapes, ranging from small gardens to old-growth forests. There are countless microorganisms continuously performing different functions that help to maintain healthy ecosystems in the soil. When the biology is improved in the soil, the vital ecosystem services that are largely used by society are promoted The Food Web in soil has great.

Some of the importance of soil food web includes;

- Disease suppression - Soil that is plenty with a diversity of microorganisms decreases the viability of human pathogens and plant in the soil.
- Decomposition of organic matter - The major decomposers in the soil are bacteria and fungi. Animals that live in the soil such as mites, millipedes, earthworms and termites assist in the decomposition process by dispersing microbes throughout the soil and tearing the organic material. These groups of decomposers play a vital role in controlling pollution and waste management.
- Nutrients retention - Loss of nutrients in the soil due to the movement of water down through the soil is known as nutrient leaching. This can be prevented by bacteria and fungi bystoring the nutrients in the soil in their bodies. Therefore, nutrient leaching is mostly resisted by soil that is rich in these organisms.
- Cycling of nutrients - The fecal excretions of predators such as protozoa from consuming bacteria and fungi release nutrients that plants then absorb when the predators exist in the soil. These nutrients help in flourishing of plants. There is elimination of any need for chemical fertilizers in the presence of these predators.
- Disease restrain - Soil that has variety of different microorganisms reduces human pathogens and the viability of plants in the soil.
- Toxin decomposition- Contaminants get degraded by fungi decomposers. In aerobic conditions, decomposers such as bacteria can break down pollutants and pesticides in soil. Fungi degrade and sequester contaminants in the environment.

7.4 WAYS TO ENHANCE SOIL HEALTH AND SUSTAINING SOIL-BORNE PATHOGENS

7.4.1 Minimum Tillage Practices

Minimum tillage is described as any method of tillage which reduces the amount of tillage passes, such as that soil aggregate disturbance is minimized, and a bare fraction of 30% of the ground surface spread with residues that help to curb soil erosion. This procedure decreases debrisburial, including refuseretention during sowing or addition of stubble mulches which aid in covering the ground.

When expertly used, it can help to control the propagation of aphid vectors, thereby decreasing virus spread as well. Similar to plant canopy cover, minimum tillage works by lessening exposure of bare soil between crops and thus deterring incoming aphids. These are significant control factors in the initial stages of crop growth, just before the canopy top shelters the soil surface in-between rows. The tillage practice is also especially helpful when crop stands are narrow and row spacing is wider (Patel, 2016).

Figure 7.7: A farm where minimum tillage is practiced as one of the ways to improve soil vitality.

Source: https://www.greenlife.co.ke/11-advantages-of-minimum-tillage-zero-tillage-systems/

Benefits: Minimum tillage minimizes soil disturbance and decay, thus providing surface residue protection that can improve water-retention, soil nitrogen and carbon, plus potentially farm yield. Additionally, changing tillage affects residue placement including ground's micro-climate, by interacting with soil properties, which consequently can affect nitrate emissions.

The practice also leads to improved soil fertility. Through the actions of reduced soil disruptions, enhanced organic matter and rich organic soil matter, your farm's fertility will also significantly improve, which results in healthier crops for better yields. It further allows plant matter to remain on the soilsurface, where they rot and contribute to the organic composition of the soil.

When the remnants of the preceding plants are left to decay and mix within the soil, any nutrients they had gathered is recycled back into the soil. This reduced soil disruption also ensures that tiny organisms present in the soil flourish and multiply, plus their activities of integrating compounds within the soil areenriched. The tillage method prevents soil evaporation by providing a natural cover against heat from the environment. Organic material and plant residue cover the soil thus ensuring little to no moisture loss. This is particularly useful in dry weather conditions, where water is scarce and the plant needs to make do with whatever is available. If rains are inadequate, the plant will continue maturing as a result of the extra moisture content (Patel, 2016).

Easier weed control is yet another benefit of minimum tillage. After crop planting, the weeds which sprout are effortlessly controlled with a non-selective herbicide. Following several planting sessions, the soil cover will gradually increase and moderate temperatures in the soil. In such conditions, weeds won't grow apart from the germination-line where the soil is directly exposed to sunshine.

Moreover, studies show that crops are grown immediately the rains begin germinate and develop much faster, in comparison to those where tilling is performed after the rains. Particularly due to the additional humidity of the soil. (Patel, 2016).

Farmers stand to enjoy increased profits since all the resources saved in planting operations will ultimately reduce the cost of production resulting in greater profits. It's also economical since the costs of buying fuel for tilling machine or hiring manpower for farm work will significantly be brought down.

7.4.2 Use Plant Diversity to Increase Soil Diversity

Increases in crop diversity, be it in soil space or duration, are highly likely to prompt an improvement in species variety of soil organisms through a better-diversified litter, rooting patterns, exudates and crop associations. Proper managing of the cultivated landscape instead of farming fields per se, is just as significant as hedges, field margins and so on. Plant diversification provides a crucial basis for helpful soil organisms to thrive and recolonize disturbed regions, including themselves offering a diversity of habitats for a broader range of microorganisms. Some recommended practices for plant diversity include:

- Introducing a diversity of seed mixes to the field, examples being, deep rooting crops and herbs, including species combinations for whole-crop silage

- Inclusion of green composts into crop cycles
- Use of locally modified cycles with grass-clover leys

Generally, cover crops, organic manures and grass leys offer opportunities for enhanced organic matter inputs within the soil. In arable cycles, the distribution of crop rotations in comparison to monoculture or nominal break crops has significant benefits on soil health. Nevertheless, research is still needed to come up with specific advice concerning the most suitable green manures necessary for soil/rotation combination. This includes the most fitting management method of them, such as grazing regimes or cutting. Even so, green manures have been applied in fields for several years, such that the average farmer's comprehension of practical combinations is growing (Patel, 2016).

For livestock and mixed farming methods, diversification within silage plants and in foraged swards could be a critical tool in improving on-farm crop variety. Whole-crop silage, furthermore, is growing in capacity and plant species combinations are increasingly becoming common. In some farms, legumes are also applied to improve forage protein amounts and minimize nitrogen fertilizer necessities. Diversified species swards may solve a number of farming objectives, such as livestock health improvement and fertilizer reduction. Therefore, there's a broad range of farming benefits that come with the implementation of plant diversity.

Soil Advantages:

- Synthesis: The main point of diversifying farming ecosystems is to sustainably intensify yield while decreasing typical inputs or optimizing both production and ecosystem services. Throughout the coming few years, as monoculture harvests continue decelerating for most crops, and ecosystem demands for services escalate, diversification might become a vital tool for preserving yield and ecological services in farmlands.
- Plant diversity potentially offers a partial or full substitute for various expensive agricultural inputs, including imported pollinators, pesticides, fertilizers and even irrigation. Variation strategies include improving plant genetic diversity, blended planting, crop rotation, agroforestry and diversification of terrain surrounding croplands.
- Weed suppression: Diversification of plant systems in temporal and these measures can effectively regulate weed populations. Furthermore, crop diversification can be done by sowing a plant in

a cyclic sequence alongside other flora. Diverse rotations typically use crops with varying planting and harvest seasons, contrasting growth patterns and competitive characteristics. Plus, dissimilar tillage, or weed control practices. Weeds can easily be challenged using environmental stresses and mortality possibilities, further given limited consistent chances for unrestrained growth and reproduction.

There's strong evidence to show that strategically improving plant diversity can increase crop yield, harvest stability, pollinators, as well as suppress weed and pests.

7.4.3 Avoidance of Compaction through Traffic Control

Soil compaction is a key issue in modern farming, which can affect soil physical and molecular properties including soil-based processes. Compressed soil is economically unsustainable to farming efficiency since it results in minimized harvest yield and greater fertilizer and energy input needs.

Figure 7.8: Soil compaction.

Source: https://www.lfu.bayern.de/boden/boden_gefaehrden/bodenverdichtung/index.htm

Common causes of compaction: Soil may become compressed when exposed to high equipment traffic, or from excessive animal trampling.

Nevertheless, livestock effects on compaction are more prevalent in field systems, compared to those caused by machinery, since this is localized. The compaction effects from livestock further depend on trampling intensity, though where this is pronounced it may affect soil-bulk compactness, penetration resistance, hydraulic conductivity and macropore capacity.

Some components that determine the extent to which soil may be affected include; texture, organic matter, and water quotient. The amount of organic matter present in soil shall also influence the extent to which it's compressed, by improving its resistance to external pressure or enhancing elasticity potential. Usually, soils with great humidity content are more susceptible to the effect of compaction, since the soil strength is highly dependent on moisture content. Therefore, timing machinery utility or livestock grazing are effective strategies for minimizing the potential for compression (Schonbeck& Stein, 2014).

Solving the problem: Even when left alone, the soil will still recover from compression through natural procedures, like soil microbe activity or periodic wet–dry or ice-thaw cycles. They can disintegrate soil, though this method to recovery is a bit slow and may last up to 10 years or more, depending on intensity of compaction and specific soil type. Wheeled vehicles usually disturb soil and form deeper ruts compared to tracked vehicles, because of their reduced contact area. Conversely, tracked vehicles have the capacity to decrease rut-depth by 40 percent or more, as compared to other broader or soft tire options. This despite increasing the vehicle's mass or trailer by around 10 to 12 percent. The influence of wheeled vehicles may be diminished by using low-ground pressure tires, in which case tire size is enhanced and pressures minimized to the degree of 30-120 kPa.

Additionally, modifying machinery can significantly diminish the impact, though generally, the extent of equipment traffic is the foremost factor determining the compaction rate. There are various strategies that have been developed to reduce the portion of land compacted by machinery usage, within a system known as Controlled Traffic Farming (Mark Schonbeck& Michael Stein, 2014).

This technique utilizes stable traffic lanes, which are formulated as the only sections of the field which are exposed to the farm machine's tires. It significantly moderates the effects of soil compaction throughout the entire field. Basically, Controlled Traffic Farming can be achieved through operational planning for farm systems, plus it may be aided with the use of accurate steering tools and auto-navigation technology. Lastly, increasing organic material levels within the soil and decreasing tillage concentration and frequency is also helpful in reducing compaction. Biological material content may further improve drainage, minimizing the negative effects of damp soil on compaction. Decreasing tillage further cuts back the frequency and concentration of mechanical traffic, which might moderate organic material loss from farm soils. Nevertheless, avoidance of compaction is always a better plan than remediation.

7.4.4 Managing Nutrients in the Soil

Figure 7.9: Nutrient balance in the soil leads to healthy plants which produce a good yield.

Source: http://managingnutrients.blogspot.com/2015/12/year-of-soils-4r-nutrient-stewardship.html

Nitrogen (N) Management: Nitrogen plays a significant role in plant health and growth. Successful nitrogen management is essential in optimizing plant yields and improving profitability while reducing nitrogen losses within the environment. Nevertheless, nitrogen control is unique and maybe a multifaceted task. Nitrogen is crucial in the formation of hormones, amino acids, nucleotides, chlorophyll, nucleic acids and enzymes. The compound promotes rapid crop growth and enhances grain yield, including quality through improved tillering, grain formation, filling, leaf area growth and protein synthesis.

There are several strategies that can be applied in Nitrogen management:

- Timing: Among the main challenges that most farmers face when determining a nitrogen fertility plan is the application timing. For fertigation systems, the most ideal practice is putting frequent minute applications, at rates thatmeet the plant requirements.But in less intensive plants, like grains and cereals, where just a few fertilizer applications are done, timing of the nutrient's application is key.Putting nitrogen too early increases the danger of losing it from leaching, before the plant even utilizes it, particularly if rains are coming. The best approach, in such scenarios is subdividing the nitrogen application, whereby majority of the nitrogen-fertilizer is applied just before the plant's growth stage where maximum nitrogen demand is necessary. Even so, there's still a risk of putting the nitrogen fertilizer very late, if logistical or climatic conditions don't permit for spreading it when planned.
- Managing application rates: Nitrogen undergoes rapid and constant changes once introduced in the soil, leading to different forms which are quite mobile. Therefore, testing soil nitrogen will provide a reading which is valid at just the moment of testing, but might result ininaccurate recommendations for nitrogen utilization. And so, the standard approach is giving nitrogen recommendations depending on yield goal, as well as nitrogen uptake levels of the plant. Nitrogen credits, as a result of organic matter within the soil and remains of previous crops, must be taken into consideration when giving nitrogen fertilizer rate advice. Also, research is ongoing regarding new methodologies and techniques for measuring soil nitrogen which scientists are presently developing and evaluating.

Atmospheric nitrogen forms a primary reservoir of the compound in the environment, even though it's unavailable to many crops. Only legumes have the ability to utilize atmospheric nitrogen in organic processes involving bacteria, while negligible amounts of nitrogen may also be deposited by rain (Schonbeck& Stein, 2014).

On the contrary, most nitrogen present in the soil is available as organic matter, which is reasonably stable but not directly accessible to florae. Plants may absorb nitrogen just in its mineral forms of nitrate and ammonium. It's only about 2-3% of nitrogen found in organic material that becomes accessible to plants each year, through a process known as "mineralization".

This procedure involves bacteria thatchange organic nitrogen to its mineral form, which is readily available to crops. The mineralization method is determined by environmental factors, including temperature, soil pH, humidity and aeration. Case in point, excess moisture restricts the availability of this compound, while also slowing down mineralization process. This procedure is usually optimal at 30°C and pH levels that are neutral to slightly high.

Phosphorus (P) Management: Phosphorus is among the top 17 essential nutrients needed for crop growth. Despite its significance, the compound limits plant yield on around 40% of the globe's arable land. Besides, studies show that global reserves of the compound are being diminished at a fast rate that by 2050 according to various estimates there might not be any amount left in the ecosystem. To ensure proper utilization, residue contained in plant remains if still available in the farm can be reused, by integrating the residues back into the soil where it would provide nutrients. Similarly, crop residues given to livestock may be recycled back into the soil as organic manure or bone meal. Other uptake efficiency methods for phosphorous include; adaptation of root architecture, creation of larger root systems and application of low molecular mass organic acids (Schonbeck& Stein, 2014).

The concentration of phosphorus in plant yields involves more than only its availability in the soil. Other factors that play a role are; soil pH, crop root growth, method of phosphorus supplementation and the other similar factors which influence root growth.

- Soil Test: Soil test is among the most crucial tools for this mineral's management in crops. It shows the soil's pH level, phosphorus quotient, and defines the appropriate application quantity of phosphorus that should be applied for adequate crop growth. Regular and representative soil-sampling technique is crucial for accurate analysis of soil test results. Likewise, sampling depth is considerably important for pH and phosphorus, particularly in minimal and no-tillage methods where there's negligible or no blending necessary to standardize the soil. There is not any specific "available" portion of phosphorus present in soils. What's available occurs in solutions, including future projections of mineral solubility and organic material over the growing period. As a result, soil tests can't extract the specific available quantity from the soil, instead only an amount thatreflects what could become available.

- Application method: Due to this compound's immobility plus soil fixation properties, adding of fertilizer phosphorus may impact on its availability to crops. Fertilizer that's broadcast and tilled down is usually assorted uniformly using considerable amounts of soil. Therefore, the potential of root contact together with the manure is greatly enhanced. Similarly, though, added fertilizer could be in greater contact with absorbent surfaces in the soil, thus escalating phosphorus fixation. Once the fertilizer is spread likea highly concentrated band, surface contact and fixation would be minimized. Nevertheless, lack of phosphorus dispersion from the area of placement further means the numberof roots that are in touch with the fertilizer would be less, compared to broadcast or plowed down methods. The higher the capacity of soil to assimilate phosphorus, the elevated the significance in superseding the fixation capacity using a concentrated band. Crop response to phosphorus placement, however, may be determined by crop root features, soil phosphorus amounts, and temperature levels. For soils with moderate up to high amounts of phosphorus, banding method of application may have lesser advantage, whereas broadcast applications are mostly adequate plus sometimes better than banding. It's also been found that nutrient intake of surface-spread fertilizer is equivalent to or surpasses uptake under typical till management methods.

7.4.5 Cover Crops Incorporated as Green Manures

Compared to artificial fertilizers, green manures are cheaper. Green manures help in giving out high yields in a cheaper way. Green manures can be used in place of animal manures. To grow a green manure, one digs the soil when the plants are still young before the plant's flowers are produced or before they produce any product. Green manure is created after withering of the plant parts to the soil and they act as mulches. Cover crops are generally green manure and green manure ismainly grown to function as cover crops. Their growth is very quick and within short time the biomass builds up. Green manures help in increasing the rate of productivity and conserving the environment (Idowu, 2014).

Figure 7.10: Cover crops being turned to green manure.

Source: https://www.flickr.com/photos/145154178@N08/39346662591

Cover crops are fused into the soil when tilling when they start flowering or when they are green. The following are some of the major advantages of cover crops incorporated as green manures:

- They give shade to plants that why it has the name cover. Cover crops acting as green manure prevent the soil from drying and heating up. They also prevent the soil from being carried away by wind hence it prevents soil erosion. When fused to the soil green manure cover crops acts as the cause of the organic matter.
- There is maintaining the fertility of the soil by the soil organic matter. Normal fertilizers cannot maintain soil fertility as the organic matter. Nitrogen from fertilizers is not needed since most of these cover crops include the legume crops since they are able to absorb the nitrogen gas into soil. They also help prevent crop pests and diseases which negatively affect crops grown especially during mono-cropping in farms.
- Cover crops that act as green manure also help in avoiding or preventing leaching of soil nutrients from the soil in hot areas. This occurs when the major grown crop has been harvested and the land remains bare.

- These green manure cover crops increase organic matter since there is an increase in biomass compactness which speeds soil aggregation and the activity of microbial. Cover crops should be planted when the soil is not wet.
- There is slow loss of nutrients- increased biomass decomposition leads to the nutrients being released slowing from the soil hence the nutrients are fully utilized and nitrogen is given to major crops.
- Cover crops also reduce soil pathogen and weeds this is by smoothening the weeds. They also decrease the moisture quantity from soils that are wet there is the reduction of the soil from being waterlogged.
- Some of the cover crops that are green manure provide a feed to the insects that pollinate flowers hence they provide shelter to the insects. Theses decrease the chemical use of insecticides where there are cover crops.
- Incorporating cover crops in soil assist the nutrients in the green manure to be ready and released to the crops.
- Green manures play the role of soil acidification of matter to reduce the alkali level.
- Time, energy and resource are the limits to consider while using green manure. These are the factors required for thesuccessful growth of the cover crops.

7.4.6 Mulching

Mulching is the process of covering a bald land or farm soil to increase soil health. If done well, mulching is important to plants. It is done by using mulches to cover the soil. Some organic material, leaves and woodchips are examples of the mulches used. These are used to prevent moisture loss and to enhance good soil conditions. Mulching saves time in garden maintenance. The waste from the farm can be used as mulches (Idowu, 2014). These mulches also prevent the growthof weeds.

Mulches are applied to the period when the soil is warm hence it is not advisable to apply mulches when soil is wet or so dry.

Importance of mulching:

- Mulching prevents weeds from growing this is when soil is fully covered by mulches hence preventing the seeds of the weed from germinating since they will not get the light from the sun that is necessary for their growth. By covering the whole farm or garden with mulches the weeds will not be able to be attached to the soil hence making it easy to pull out weeds.

- During mulching, there is retention of moisture in the soil for good plant growth. A plant likes to grow well when the amount of moisture in the soil does not reduce hence the plant will be healthy because diseases together with pests will not likely affect it. In addition, the soil is kept cool by mulching and this favors plant growth.
- Mulching adds nutrients to the soil. Mulch is also rich in organic matter and this is also added to the soil. These organic matters are decomposed and then microbes and earthworms play the role of breaking down organic matter. This enables plant to have healthy growth.
- Mulching changes the soil pH with time.

Mulching is categorized into two main groups: inorganic and organic mulching.

Organic mulching

The following are some of the organic mulching materials:

- Leaves- when leaves fall from trees mulch is naturally formed.
- Hay, helps in keeping the soil cooler
- Clippings of grass- The cut grass are used for mulching. Seedlings can be mulched by these grasses. These grasses also give the soil nitrogen when they decompose.

Disadvantages of organic mulching:

- Pests may find a habitat to stay in and this has effects ongrowing plants.
- The mulches may have some weed seeds.

Inorganic mulching

People use inorganic mulch due to lack of enough organic mulch. These includes:

- Using black mulches which prevents light from entering the soil since light is absorbed by the black material. This prevents weed growth since light cannot reach the weed.

- Using transparent mulches which allows the passage of light and encourages weed growth usually practiced in nurseries.

Figure 7.11: Mulching in strawberries garden.

Source: https://www.palmers.co.nz/faqs-strawberries/

7.4.7 Composts

Composts aredone through composting which involves decomposing of organic materials to nutrients in the soil by organisms in the soil. Compost helps in improving and re-energizing the fertility of the soil. It does not require use of fertilizers. All organic materials are compost materials and nutrients are added to the soil by composting (Patel, 2016).

Figure 7.12: Composting.

Source: https://sustain.olemiss.edu/composting-program/

Compost soil is good for productivity in agriculture. Compost has organic matter which increases the biological property of soil. These help as follows:

- the soil to avoid diseases and pests,
- Enables the transport of the nutrients to various parts of a plant.
- The microbial activity increases the cycling of nutrients in the soil.

Types of composting

Composting is grouped into two;

- Anaerobic composting- happen when there is little or no oxygen. Organic acids, methane and other substances are developed by anaerobic small organisms. The substances have a bad smell. There is no metabolization in these substances when there is no oxygen. Anaerobic composting is favorable in low temperatures and these leaves the weeds not damaged. Anaerobic composting takes more time than aerobic composting.
- Aerobic composting- This one takes place when there is enough oxygen. There is the production of water, ammonia, heat, humus and carbon dioxide when the organic materials are broken down by the aerobe organisms. The aerobe microorganisms decompose substances that are produced such as organic acids. The heat that is produced helps in breaking down fats and proteins. Aerobic composting leaves pathogens and weed damaged when there are high temperatures. Aerobic composting is a quicker process than anaerobic composting. Its limitation is that there is loss of many nutrients. Aerobic composting is however preferred to anaerobic composting.

Factors that affect aerobic composting

- Temperature-a high temperature of 50 to 70 □ is involved in this process and this helps in damaging pathogens and also completely destroys weeds.
- Soil moisture-Moisture plays a major role in metabolism. A moisture amount of about 60% is needed for the materials to compost.
- The value of the pH - Eight should be the pH value of composting material. There may be ammonia as a loss if the pH level is higher than eight.

- Soil aeration- Since aeration contains a lot of oxygen gas, aerobic composting requires needs ample oxygen. Microorganisms require enough oxygen to grow. Aeration decreases the amount of water, vapor and heat in the soil.
- Nutrients in the soil- Nutrients such as phosphorus, and potassium are required to compost materials.

Application for proper aerobic composting

- Improving aeration- the soil should receive enough oxygen to get a good quality of compost. And this enables the microorganisms to play their role well. Ventilation can also be put in the piles by putting up holes.
- Shredding of materials- This increases the surface area for microbial activity by cutting of the materials such as the wood to speed up aeration.
- When lime is added the plant, materials structures become weaker to increase the population of microbial.

7.4.8 Reduce Pesticide Use

This involves a decrease in the use of chemicals to kill pests in soil. This enables the soil to have good health. There are different techniques to reduce the use of pesticides. Bringing up agricultural techniques that do not allow disease, pests and weed to reach a point of destroying crops. Growing of the crops that resist pests, diseases and weeds and also practicing agronomic (Patel, 2016).

The agronomic practices: Agronomic activities facilitate healthy growth of plants and preventing weeds, pests and diseases from affecting crops. These practices include;

- Good timing of planting of crops to reduce the pest effects.
- Doing proper irrigation so that to prevent the soil or plant from having more or less water hence reducing pests and the presence of weeds.
- Practicing crop rotation by planting different crops every season to prevent pests and diseases from being on the farm for seasons.
- Through practicing intercropping- this is the planting of more than one crop in a garden. This is important in reducing the spread of pest and diseases since intercropping brings up their repellant.

- Practicing shallow plowing to prevent or reduce weed growth in every season.

Using natural pesticides: There are natural materials that can be used to resist pests and diseases together with weeds and reduces pests and weeds from reproduction.

Use of resistant crops: Using unique crops that are good for the preventing and managing of weed, pests and diseases. Using the resistant crops and doing rotational cropping. Resistant crops can be built up by generic engineering and by cross-breeding.

Other strategies for pesticides management:

- By checking the population of pests in the garden or farm, avoiding the use of pesticides sometimes, timing the stages of cycle of some pests to use the pesticides.
- Practicing the integrated management of pests – This is the process whereby all the important techniques of managing pests without the use of pesticides are brought together. The importance of this is to decrease the use of pesticides and using good chemicals thatare not harmful soil, organisms or plants. Integrated pest management helps in controlling pests for a long period of time. It also leads to high production.
- There is also the use of culture control to prevent the losses in crops. This helps in damaging and resisting pests from affecting crops. The cultural methods include; rotation of crops, timely harvest, timely planting, using the necessary seeds, intercropping. Use of organic mulching and manure.
- Weeding by the farmer is also important in controlling pests in gardens, tilling of land and use of hoes included. Weeds are covered by soil or buried when the garden is being tilled or plowed.
- Using physical or machine controls such as using the light to attract and catch beetles and some other pests and also using machines that helps in the pruning of the weeds.

7.4.9 Crop Rotation

This is the practicing or process of growing different types of plants in each season in the land or garden. This is practiced to increase the fertility of the soil and crop production. There is reduction of the fertility of soil and the yielding of a crop if only one crop is grown in a particular piece of land in every season. This is known as mono-cropping (Idowu, 2014). Crop rotation reduces the population of pests on a farm that would increase if mono-cropping is practiced. Crop rotation also helps in adding nutrients that weretaken away by the crop grown in the previous seasons. In addition, there is increase in biomass and the fertility of soil and the structure of soil. Land can likely be eroded if only one type of crop is grown repeatedly each season hence there is need for crop rotation which helps reduce soil erosion. It also makes the soil stable.

Things to consider when doing crop rotation are:

- Cereal crops when include in crop rotation improve crop production and soil fertility.
- There is a need to consider the amount of nitrogen in the soil nutrients when considering crop rotation. This is done through testing of the soil. Legume crops help increase nitrogen into the soil and hence there is less need for fertilizer usage.
- The amount of water and where the nutrients are also influence crop rotation. There is more accessibility to the nutrients and water that are deep down in the soil when the crop has a deep root. Examples of these crops are the cereal crops while crops such as the pea have roots that are shallow.
- The previous type of diseases and time that you grew a certain crop-crop rotation will help in controlling the diseases and helps in decreasing the number of pathogens
- Considering weeds that might appear – this is vital since one can know how to manage the weeds and weeds that might emerge.
- Soil types- One should know the type of soil of the land to practice crop rotation whether it is fertile or infertile.
- Kind of irrigation to use – there are different types of irrigation to every crop in the crop rotation.
- Insects, pests and diseases influence the crop rotation. The yield of a crop decreases if it is planted repeatedly every season in the same land due to increasein damage by pests.

7.5 ROLE OF CARBON IN MAINTAINING SOIL HEALTH AND SUSTAINING THE SOIL BIOLOGICAL COMMUNITY

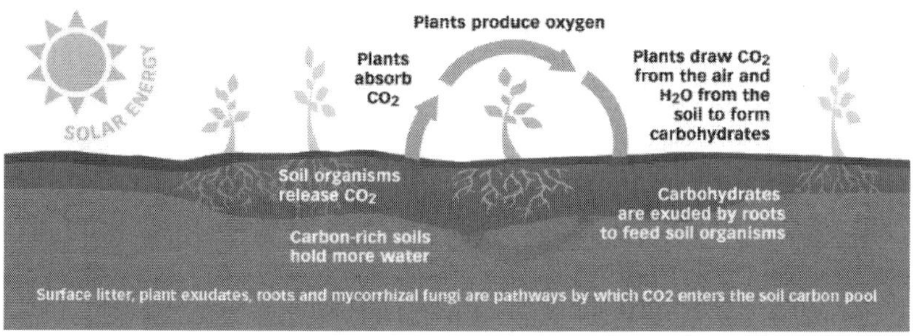

Figure 7.13: Carbon is the building block of all organic matter and plays a crucial role in soil preservation.

Source: https://www.cecsb.org/rethink-food/carbon-farming/

Carbon stored within soil organic matter is essential in promoting soil health, vitality and ecosystem services, such as food production, hence making its conservation and restoration necessary for sustainable development. Even as climate change occurs, with adverse impacts on crop cultivation, farmers still need to generate around 60% more food before 2050 as the human population continues to swell. To achieve these needs, soils must be highly productive and one of the ways to do this is balancing the carbon level.

Research shows that soils containing a high carbon ratio are more likely become productive for the farmer, plus filter and cleanse water passing through it. The water present in soil acts as the main source for around 90% of the globe's agricultural production, and further represents 65% of the globe's freshwater reserves. At increased depths, soil tends to hold thrice as much carbon compared to the atmosphere. Likewise, soil organic-carbon is usually sensitive based on management practices. Unconstructive land management practices, for instance, may make soils lose their carbon matter and emit greenhouse fumes as a result (Idowu, 2014).

7.5.1 Soil Management Techniques for Preserving and Improving Organic Carbon

Adopting agrarian practices that preserve soil organic carbon has been found to increase the yields of wheat, soybean and maize. In fact, a single ton increment in the soil's organic carbon mix of degraded cropland might improve wheat yields by around 20 to 40 kilos per hectare, that of cowpeas by 0.5 to 1 kilos and maize by 10 to 20 kilos.

Moreover, cover crops applied as green manure play a crucial role in management for purposes of increasing carbon stocks. Agroforestry Systems, on their part, stock carbon in the trees in form of woody biomass which reduces greenhouse fume emissions from soils. Studies reveal that agroforestry biospheres store around 7 tons of carbon for each hectare per year. Furthermore, combined application of compost and chemical fertilizer can cause up to 43 percent increased soil organic carbon storage, in comparison to inorganic fertilizer use in arable lands (Idowu, 2014).

7.5.2 Impact on Biological Communities

Life itself is built upon the conversion of CO_2 into carbon-based organic elements that make up living organisms. This carbon cycle demonstrates the central significance of carbon within the biosphere. Biological cycling of this compound starts as photosynthetic cells integrate carbonates or carbon-dioxide from the environment. Within terrestrial communities, crops convert CO_2 in the atmosphere to carbon-based elements through photosynthesis. In this process, crops hold onto the carbon compound from a pair of oxygen molecules, then release the oxygen again into the environment.

Plants are, therefore, predominantly responsible for regulating atmospheric oxygen. In aquatic farming activities, like growing rice, crops use dissolved carbon within the water to form carbonates or carbon dioxide, which acts as the primary supply of carbon for photosynthesis. Moreover, once carbon is fully assimilated with the help of photosynthetic organisms, including the animals which feed on them, it's released back into the ecosystem as carbon dioxide when organisms respire. This release of CO_2 back into the atmosphere marks the completion of the biological phase of the carboncycle. Nevertheless, different carbon cycle paths will recycle the compound at varying rates (Idowu, 2014).

Chapter 8

How to Control Harmful Soil-Borne Pathogens

CONTENTS

8.1 INTRODUCTION

A pathogen is a microorganism that can cause disease. The soil is the upper part of the earth in which plants grow. Therefore, soil-borne pathogens are microorganisms that live and survive in the soil that may cause disease in plants, especially in the roots. They may also affect the soil. These pathogens can be very destructive and are one of the prominent limiting factors to farmers' income. Soil-borne pathogens include fungi, bacteria, viruses and nematodes. Examples of these pathogens are Fusarium spp, Verticillium spp, Rhizoctonia spp and Phytophthora spp. These are just but a few. Just like any other microorganism, they require certain conditions to survive and thrive. These conditions could be environmental, such as temperature, pH, moisture, adjacent surrounding, adjacent surroundings, etc. They have different life cycles, therefore one has to note and understand the weather conditions, soil conditions, location and other factors that would have direct impact on the diseased plant, such as weeds or even plant density.

These pathogens cause soil-borne diseases that may be characterized by rotting of roots, rotting of the stem, damping-off diseases, stunting, yellowing and spotting of lower leaves, wilt diseases, etc. These diseases cause many negative effects such as; reduced plant growth, increased costs to the growers and even damage to the environment.

Figure 8.1: This shows a pathogen has affected the maize crop by root rot rendering the maize wastedcausing loss of income for the farmer.

Source: https://www.grainsa.co.za/new-prospects-for-tackling-the-problem-of-soil-borne-diseases-of-maize

8.2 CULTURAL CONTROL METHODS

Cultural control is the modification of the plant growing environment to reduce the prevalence of pathogens. This may happen by either directly affecting their growth or reducing the chances of the pathogen attacking the plant. One must have knowledge of pest ecology and large-scale adoption. Under this topic, we discuss 8 main methods; fertilizer application, flooding, drainage, rouging, tillage practices, plant density, weed control and growing of cover crops.

8.2.1 Fertilizer Application

Fertilizer is a chemical or natural substance added to soil to increase its fertility. Mineral nutrition is important in crop production but it can also promote disease development. Excess or deficiency of minerals can make the plant susceptible to soil-borne diseases. These are abiotic disorders. Minerals can cause specific ion effects or total ion effects having effects on the pathogen or the host itself (Sullivan, 2004). It is true to say that specific ions could harm the pathogen, specific ions could harm the plant, ions change the pH of the soil making it more or less suitable for the pathogen or the plant, ionic strength changes the environment of the root making the plant weak and susceptible and ions change the soil environment making the plant susceptible to disease.

There are different methods of applications: spot application, banding, fertigation and dolloping.

Accumulation of Silicon in the cell walls prevents penetration of fungi due to its assistance in the formation of a protective physical barrier. Silicon also activates the plant's defense mechanism, allowing it to resist pathogens. In contrast, excessive Nitrogen levels lower the Silicon content, making the plant susceptible to pathogens such as fungi, resulting in increased disease incidence and lesion area. However, Nitrogen is not always associated with an increased disease incidence, e.g. rice and maize grow faster in nitrogen fertilizer. Copper is widely used as a fungicide, relying on the direct application on the plant surface and the infecting pathogen. Copper deficiency may lead to impaired defense and reduced lignification, contributing to lower disease resistance (Sullivan, 2004).

Figure 8.2: Excess Nitrogen in plants can lower the integrity of the cell wall, making the plant susceptible to pathogenic diseases.

Source: https://homeguides.sfgate.com/effects-much-nitrogen-plants-43755.html

All these show that one must be very thorough in choosing the correct fertilizer with specific minerals and knowing the right amounts to apply. This will in turns help to control soil-borne pathogens.

8.2.2 Flooding

Prolonged water-logging lowers oxygen levels and results in the accumulation of CO_2. This method can be used as a physical or cultural method. An example of flooding is rice paddy. It is mainly used to reduce the prevalence of weeds. It can also be used to eradicate fungi and nematodes in the soil. Weeds are known to carry disease-causing organisms and insects which may harbor viruses and bacteria. This can be the main reason why weeds are a big factor in the control of soil-borne diseases.

Some pathogens can be controlled by flooding, and some can be spread or favored by wet soil. An example that can be spread or favored is late blight (Phytophthora) and one that can be controlled is Fusarium wilt on bananas. Flooding can also hasten the destruction of crop debris carrying pathogens. This is normally not feasible as it uses a lot of water and makes the land lack purpose for a long while thus losing economic value (Sullivan, 2004).

Root-knot nematodes can be controlled using flooding by use of some anaerobic organisms but it is not cost-efficient. The pathogens cause an alteration of the plants surrounding environment preventing the nematodes from having favorable conditions to cause disease.

However, lack of other factors such as drainage may make flooding as a means of disease control useless because of a probability of the other pathogens that may survive in soil water saturation emerge. Therefore, one has to consider all the possible outcomes and ensure they have the correct conditions and soil to use flooding as a means of disease control and also to avoid production loss (Sullivan, 2004).

8.2.3 Drainage

Drainage gets rid of excess water in the soil thus reducing water saturation and allowing soils to retain their original and fertile state. Most of this water saturation occurs on the top layer of the soil, making it easier to drain. The ultimate purpose of draining the soil is to be able to ensure the soil can support plant growth.

Water saturation can cause a deficit of Oxygen in the soil, due to the spaces in the soil being filled by water. Water is known to have a lower percentage of Oxygen than air and plants cannot get enough oxygen which they need to thrive from the soil if the soil is saturated with water. This can act as an advantage when it comes to disease control, as aerobic organisms cannot survive in an area without oxygen, therefore propagating the death of aerobic pathogens, assisting in the control of soil-borne diseases (Puri, 2016).

There are different types of soils, all with different properties. Some soils with clay may have a harder time draining water due to their close-nit, small particles. This promotes flooding in such soils. Some fungi thrive in such conditions (soggy soils) and may cause root diseases.

Lack of drainage may cause loss of Nitrogen; Nitrogen is a major nutrient in reducing soil-borne diseases. Without nitrogen and other nutrients, then the disease resistance of plants is reduced due to lack of stimulated plant growth and nutrient availability, inhibiting the escape of pathogens (Puri, 2016).Loss of Nitrogen can be beneficial in allowing absorption of Silicon which acts to increase the integrity of the cell wall allowing the plant to defend itself from certain pathogens.

Poor drainage should be dealt with to prevent the above consequences. One can deal with poor drainage by increasing the amount of organic matter into the soil, or by the building of raised beds on top of the wet saturated ground. To identify diseased plants, one need to look at the color of the leaves which may be yellow or pale green, the integrity of the stem can also be observed.

8.2.4 Rouging

This is the identification and eradication of diseased plants, also called rouges thus the name rouging. It is done to avoid the continuous spread of pathogens among plants.This practice is efficient in gardens where the labor is cheap or not important or when the market value of a crop is high. This is mainly in small gardens. Roughing is only efficient in larger plantations only if the crop being grown is of high market value and the key signs of the disease are very visible and raise attention (Puri, 2016). Any diseased plants should be cut or plucked out from the root, depending on how extensive the infection is, so as to minimize the spread of the pathogens. This should be repeated regularly during the whole period if any conspicuous symptoms arise.

Figure 8.3: Physical signs such as wilting, change in color, etc. especially in plants with high economic value should be removed due to the risk of transmitting the illness to other surrounding plants.

Source: https://www.youtube.com/watch?v=Ew9jjGXebU4

To complete their life cycle, some pathogens may need alternate hosts. For example,Cronartiumribicola requires pine and currants for a full life cycle. In this case, removing the currants interrupts the life cycle of Cronartiumribicola thus leading to control of the disease. One has to check for the color and shape of the crops to ensure the integrity of the crop. Anything other than the usual should be removed to prevent the spread of the pathogen.

Rouging should be highly considered especially if one has invested a large amount of money into the crop. A special emphasis is needed during the period of panicle emergence to early seed development.

8.2.5 Tillage Practices

Tillage is the manipulation of the soil mechanically aimed at improving its physical condition, providing favorable conditions for growth. It incorporates different types of organic matter to help reduce weeds volunteer crop plants that may act as a vector to different pathogens. It also works by turning over the soil, thus putting underground the pathogens in the topsoil into deeper layers which is not a suitable place for the pathogens to cause disease (Puri, 2016).

Figure 8.4: Tillage works by burying plant pathogens from the topsoil into deeper layers of the soil causing less or no disease.

Source: https://serc.carleton.edu/integrate/teaching_materials/food_supply/student_materials/1165

When tillage is done in hot summer periods, the resting spores from the pathogens are exposed to the sun and thermo sensitive pathogens die eg, Nemawool. Tillage is used in the control of common soybean pathogens such as Sclerotina Stem Rot. Sclerotina Stem Rot has the ability to survive in the soil and even in crop debris. Seedling diseases (Fusariumand Rhizoctonia) can be controlled using tillage due to its ability to improve soil drainage and also by increasing the temperature in the soil, allowing the soil to retain moisture (Puri, 2016). Regardless of controlling soil pathogens, tillage has disadvantages such as soil erosion, soil compaction, increased agriculture runoff and it also consumes a lot of time and resources.

8.2.6 Plant Density

Crop density is characterized by the planting of crops close to each other in aim to increase the yield. It can also be done to prevent transmission of some diseases as will be seen below. It has shown a huge impact on the topic of control of soil-borne diseases. Planting of crops close to each other can be a major way for crops to pass on pathogens from one to another. Plants in close proximity, exhibits higher chance of passing on pathogens from one to another due to the reduction in the distance that the pathogens or their vectors have to travel being reduced or even because the leaves of the crops have high chances of coming in contact.Crops being in close contact can also cause an increase in humidity due to a constant high temperature causing leaves to stay wetter for longer periods of time in times of rain or dew. Humidity can favor the development of disease by allowing favorable conditions to some pathogens (Agtmaal, 2015).

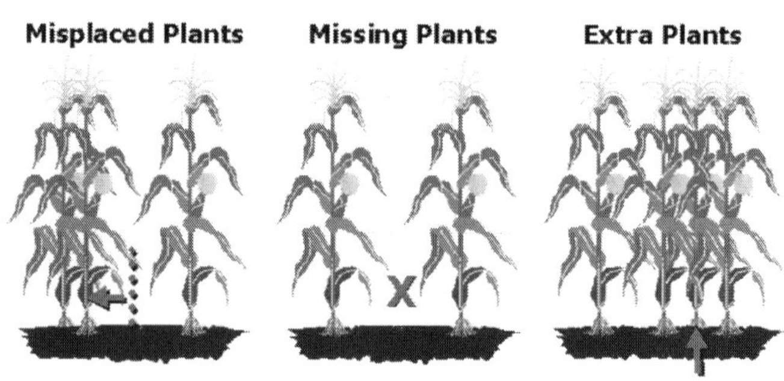

Figure 8.5: Plant density is characterized by the planting of crops close to each other so as to increase yield or even in some plants that require continuous ground cover to prevent pathogens.

Source: https://www.mycoffeepot.org/planting-depth-for-sweet-corn.html

Crops being in close contact can also cause an increase in humidity causing leaves to stay wetter for longer periods of time in times of rain ordew. Humidity can favor the development of disease. The broken ground can reduce the incidence of disease in some pathogens, such as the vector of the groundnut rosette virus. Seeing as this is the case, the weeding should be discouraged as it leaves breaks in the ground (Agtmaal, 2015).For crop density to be efficient, root contact should be avoided by methods such as pruning, staking and harvesting plants and other methods depending on the crop.

8.2.7 Weed Control

Weeds are a sanctuary for insects thatcarry many virus diseases. Seeing as most viruses require to be in a living organism to remain viable, destroying weeds in and around the garden allows in the prevention of soil-borne diseases by eradication of any disease-carrying insects.Management of weeds is important in production techniques due to their extensiveness in plantations but is expensive and hard to achieve. There are four methods of weed control: mechanical, cultural, chemical and biological. Here we will discussabout the cultural methods in detail (Puri, 2016).

One way of weed control is selective crop simulation. This can be achieved by the application of soil amendments like lime which may correct the soil conditions in favor of crop growth. Also, inoculation of crop seeds with suitable nitrogen-fixing and phosphorus solubilizing organisms may help in selective stimulation of some crops. Proper planting methods may also achieve weed control. For example, minimum tillage and furrow planting have proven to reduce weed growth. Growing of intercrops has shown to suppress weeds providing a chance to use crops themselves as a method of weed control. For example, green grams suffocate weeds without reducing the yield of the main crop. All these methods control weeds thus control potential pathogens that may be harbored in weeds.

Some weeds may have some advantages such as acting as manure, as fuel, as a human feed or even as soil binder.

8.2.8 Growing of Cover Crops

This is when you grow plants that cannot act as hosts to the existing pathogens on the field.For this to work well, one must understand the pathogen's stages of life, such as its soil longevity, its additional susceptible hosts, its transmission and methods of control the pathogen.These crops can control disease-causing organisms like Gaeumannomycesgraminis and some disease-causing bacteria that only survive in presence of a particular host.It is mostly used in the control of pathogens in cereals. These often include legumes that can hot the pathogen, in turn enhancing soil fertility and lowering the fallow period. Some cover crops may inhibit the development of certain disease-causing organism. For example, root-knot nematodes highly occur in soils following maize, while crops like watermelon show the presence of lesser number of nematodes (Puri, 2016).

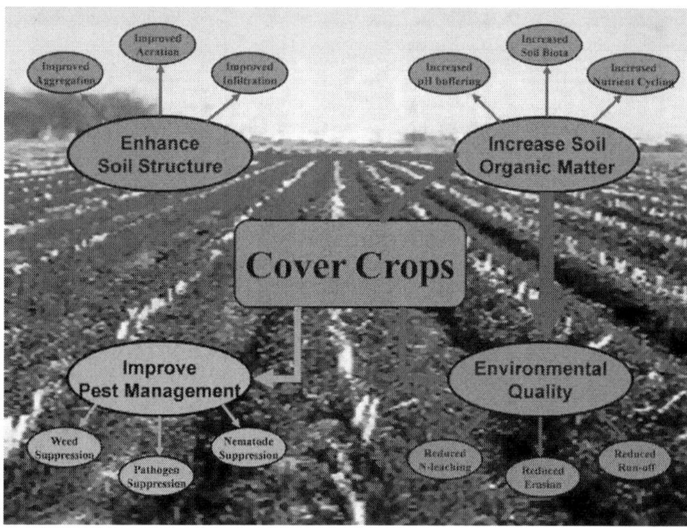

Figure 8.6: Cover crops are non-host plants, meaning they cannot host the available pathogen. This allows time for the pathogen to die without affecting the growth of the main crop.

Source: https://www.middlefieldpost.com/2017/03/28/using-cover-crops-to-increase-yield/

8.2.9 Crop Rotation

This is the art of growing different crops in a sequence on the same piece of land over a specified period of time so as to get maximum benefit with the least investment and without wearing out the soil and environment. It may include 2-6 crops, each with either a financial benefit or environmental benefit. Crop rotation is used for certain reason;

- To achieve better soil fertility and its physical, chemical and biological properties through the addition of organic matters.

- To keep the soil free from pests, weeds and diseases through various manipulations.
- To achieve higher yield without extra expenditure.
- To take advantage of the available labor power and capital throughout the year.
- To use selective crops to make the best use of residual soil moisture and nutrients by selective crops.

There is a certain criterion for selecting the crops to use in this method;

- The crop should have a short duration to grow.
- The crop should have demand in the market.
- The crop should be resistant to pests and diseases.
- The crop should be adaptable to the local climatic condition.
- The crop maturity should match with the crop demand.

Figure 8.7: Crop rotation which allows keeping the soil free from pests due to rotating to non-susceptible crops.

Source: https://farmandcity.co.zw/crop-rotation/

When the same crop is grown in the same field over and over again for many years, then many pathogens build up in the soil. The pathogens could become so many, that yield losses start to become frequent and regular. Rotation prevents this by rotating to a non-susceptible crop which breaks the cycle by reducing pathogen levels. Especially for those pathogens that survive a long time in the soil as hyphae or spores without a host, e.g.Fusariumspp. (Sullivan, 2004).

Despite the fact that rotation prevents illnesses, there are some considerations that need to be taken into account before rotating into another crop. First, any plant that shares a family has higher chance of having the same problem with pests. Therefore, one should avoid using any crops from the same family to rotate with as that might not yield the desiredresults one would hope for. Rotation into the same family increases chances of disease incidence. It may cay diseases like black rot and clubroot.

This is seen especially in the Brassicaceae family. Rotating to crops put of this family reduce the incidences and allow the pathogens and pests to die off. Secondly, rotation is not fully efficient in avoiding or eradicating pathogens with a wide host range. A pest like Sclerotium rolfsii can infect many different crops. Finding a crop that is not susceptible to it is very hard. One can attempt to use grains.Lastly, some pests can survive in the soil for a very long time. They create structures that allow them to rest in the soil for even three to five years e.g. Clubroot. Fortunately, most pests that can last in the soil for long have a very narrow host range, making it easy to still plant crops (alternative) in rotation.

Despite the fact that crop rotation has proven to be advantageous, there are a few disadvantages worth mentioning, especially if the rotation has not been properly followed. First, repetitive growth of some crops in rotation for many years may diminish the long-term benefits and may cause the development of associated weeds, insects, pests and diseases. Lastly, in rotation, it is quite difficult to shift to the crop of high demand or ethnic requirements (Sullivan, 2004).

8.2.10 Soil Solarization

This is an eco-friendly method of using the sun's power to control pests such as insects, weeds and bacteria in the soil. The process involves covering the ground with a tarp which is commonly a transparent polyethylene cover to trap solar energy. The sun's energy heats the soil to temperatures that kill all the pests including mites.

Solarization has a process to achieve the desired results. One has to first the area being covered withdebris and any plants. Thereafter, the soil is to be watered until deeply wet. It is then covered with the plastic which needs to be transparent (not white or black because they do not allow enough heat to get into the soil). One must then bury the edges in the soil to trap all the heat.

The plastic should be left in place for a minimum of 4 weeks. It is best done during the hottest time of summer. Once done, the bag should be removed (Sullivan, 2004).

Figure 8.8: Solarization is achieved by placing transparent covers on soil without debris so as to achieve optimum temperature levels to eliminate pests.

Source: http://organicfarmermag.com/2019/08/organic-soil-disinfestation-methods-soil-solarization-and-biosolarization/

This method works best in soil with clay or loam in it due to their ability to hold water for longer periods of time, allowing the formation of steam daily. This steam is what is needed to kill the pests in the soil. This method works best in the hottest time of the summer and especially when the soil is covered for 4-6 weeks. The longer the soil is exposed to the heat that is generated during the process, the more unwanted organisms die. During the solarization period, one should inspect the plastic regularly for any tears or holes and seal them as soon as possible to prevent loss of heat. Present method has both advantages and disadvantages. Some of the advantages are; it is non -chemical thus not interfering with the biotic and abiotic components of the soil, it is simple and therefore sustainable for both developed and developing countries and it is less expensive than chemical methods of disinfestation (Sullivan, 2004).

Some of the limitations are; the method is dependent on a certain type of climate (summer) which can only be achieved in certain climatic regions and during limited periods of the year. In addition, the soil remains without a crop for several weeks thus no economic yield. Lastly, it cannot be used insandy areas unless one lays drip irrigation lines under the covers and waters regularly, which requires more effort.

Pets should avoid the area during the process. If they walk on the cover, they may create holes thatmay release heat and make the method ineffective. Water should also be added if necessary, during the process.

8.2.11 Soil Amendments

This is the addition of material into the soil to improve it chemical, physical or biological properties. Compost is a mixture of decaying organic matter from leaves or manure used to improve soil structure and provide nutrients to the soil. Amendments can be used to rebuild soils thatwere damaged due to improper management, to amend poor soils and make them more usable, and to maintain soils in peak conditions.

Compost is the main material added to soil to improve its properties. Others are Farmyard manure, municipal solid waste compost, vermicompost, poultry manure, green manure, concentrated organic manure, etc. Composting improves the soil quality to reap their benefits and vitality by releasing the rich nutrients in the compost into the soil. It also prevents greenhouse gas emissions by encouraging the aerobic breakdown of organic material and reduces the amount of garden and kitchen waste going to landfill. It is also beneficial in that it promotes the recycling of valuable nutrients and reduces the use of artificial fertilizer. Probably the greatest benefit of using compost as a soil mulch in sediment control arethat it has immediate effectiveness and its ability to bind and degrade specific contaminants (Puri, 2016).

For soil amendment to be considered, the soil must have some problems that need to fix by adding organic amendments. Some of these are; salinity, solidity, hardpan soil and physically poor soil. The effects of adding organic amendments are: It increases soil aggregation and aggregate stability. It lowers the pH and promotes release of cations by solubilization of $CaCO_3$ and other soil minerals. The decomposition of organic matter causes the evolution of CO_2 and other organic acids. Soil amendments may be used to improve water retention in dry course soils thatare not holding water well, and they can be added to adjust the pH of the soil to meet the needs of specific plants or to make highly acidic or alkaline soils more usable.

Organic amendments supply all the essential nutrients for plant growth. It improves soil structure, drainage, aeration, cation exchange capacity, buffering capacity, and provides a source of food for microorganisms (Puri, 2016).

Mineral nutrition can prevent plant disease by two primary resistance mechanisms. This involves the formation of mechanical barriers such as thickness of the cell and by the synthesis of natural defense compounds such as phytoalexins, antioxidants and flavonoids. This concludes that mineral nutrition can prevent soil-borne diseases (Puri, 2016).

8.2.12 Seed Treatment

This is a method of enhancing planting value of the seed by subjecting them to a compound (nutrients, chemicals, hormones), a process (wetting or drying) or vigorous energy forms (heat, radiation, electricity, magnetism). Seed treatment can improve stand quality, increase yields and increase returns on investments. Seed treatment is harmless to the seed, it remains stable for a relatively long periods of time before planting, it provides an even coating on the seed, it adheres well without giving the seed an unattractive appearance, it doesn't interfere with seed flow in planting equipment, it is inexpensive and it is registered for its intended use.

The types of fungicidal seed treatment are;

- Seed disinfestation: this kills spores, mycelia, or microorganisms on the seed surface.
- Seed disinfection: elimination of pathogen that has penetrated into the living cells of the seeds.
- Seed protection: application to protect seed from pathogens in the soil. A systemic fungicide may provide post emergence protection.

There are two forms of seed treatment; pesticidal seed treatment and physical seed treatment. In pesticidal seed treatment, there are different categories as seen below:

- Seed coating; includes coating with chemical protectant and film coating.
- Pelleting.
- Seed priming; includes Osmo-priming, halo priming, solid matrix priming.

Seed coating is the direct application of a substance, botanical or synthetic to cover the outer surface of the seed providing a mechanism to protect the seeds against pathogens and pests. Examples of chemical protection areCaptan and for film coating is Intellicoats. Pelleting involves rolling of seeds with pesticides, nutrients, fillers, and binders with the gradual addition of water followed by drying until the correct size of the pellet is obtained.

Physical treatment involves radiation, temperature, magnetic, microwave, controlled gaseous atmosphere, etc.

Figure 8.9: Seed treatment has proven to be one of the most efficient ways to prevent plant diseases.

Source: http://louisianacrops.com/2013/03/14/corn-insecticide-seed-treatment-wash-off/

Anequipment used for seed treatment is the seed treatment drum. It uses dry formulation of fungicide which includes the active ingredient and an inert matter. The advantage is that it can be handled by the farmer at any place (field or home courtyard) and the disadvantage is that it causes chemical wastage and it's also not uniform on seed surface.Another is the slurry treater. A wettable powder is suspended in water to make slurry. A known quantity of seed and slurry is dumped in the mixing chamber where they get blended.

Treated seeds must be stored separately and labeled. Leftover seeds intended for planting the following year should be stored in a dry, well-ventilated location. Percentage germination must be checked before use. Treated seed and sweeping must be handled as excess pesticide (Puri, 2016).

8.3 BIOLOGICAL CONTROL

This is a method of controlling insects, weeds and plant pathogens by using other organisms. This method employs two ways of disease suppression; by antagonistic association and by symbiotic association. In antagonistic association, biocontrol agents suppress pathogens in four ways;

- Competition: This is when the beneficial organism out competesdisease-causing plant pathogens in the search for nutrients or colonization space in specific habitats such as the root zones. Increased competition prevents pathogens from becoming established and multiplying to levels that can cause plant disease.
- Antibiotics and secretions: These are produced by some microorganisms and work by inhibiting the growth of plant pathogens.
- Predation and parasitism of plant-pathogen by biocontrol agents, where beneficial microbes use pathogens as food.
- In symbiotic association, beneficial microorganisms protect the host plant from pathogen by keeping an association with the host plant of the pathogen e.g. Mycorrhiza.

Mycorrhiza is considered to be an association between fungi and roots of higher plants. It is considered as the most beneficial root inhabiting organisms, which forms a fungal mat over the root and protects the root of the plant from the attack of soil-borne pathogens. It protects the plant by acting as a barrier for invading pathogen, by providing antagonistic chemicals, by competing with the pathogen, by increasing the nutrient uptake ability of plants and by changing the amount and type of plant root exudates. Fluorescent pseudomonas are the most frequently used bacteria for biological control and plant growth promotion, but Bacillus and Streptomycesspp. have also been commonly used (Puri, 2016).

Induced resistance and cross-protection are two mechanisms of plant 'immunity' against a pathogen. In the case of cross-protection, an organism present in the plant can protect it from the pathogen that comes into contact with the plant later. Induced resistance is a form of cross-protection, where the plant is inoculated with inactive pathogenesis, low doses of pathogens, pathogen-derived chemicals or with non-pathogen species to stimulate an immune response. This prepares the plant for an attack by pathogens, and its defense mechanisms are already activated when infection occurs. It provides protection against a wide range of pathogens across many plant species.

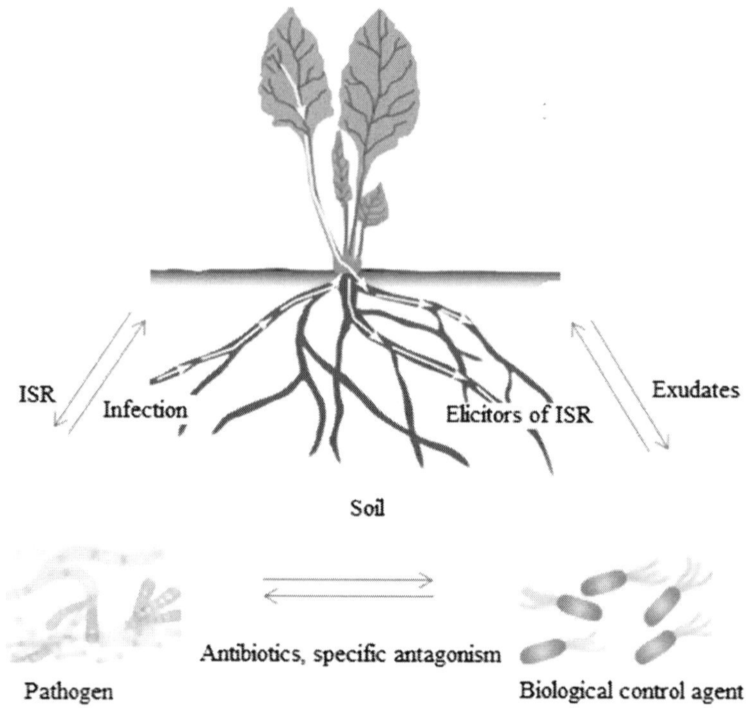

Figure 8.10: Biological control of soil-borne pathogens by fluorescent pseudomonas.

Source: https://www.researchgate.net/figure/Mechanisms-of-specific-biocontrol-agents-for-controlling-plant-pathogens_fig3_300246328

Some of the advantages of biocontrol are;

- It is harmless to human beings and animals.
- It is cheaper than pesticides by about 50%.
- It has a high cost-benefit ratio.
- It is highly effective all through the crop growth period with high rhizosphere competency.
- It is easy to deliver.
- It improves plant growth and it increases yield.
- It has the potential to be permanent, the pest is unable to develop a resistance.
- It does not intensify or create new pest problems.
- It is effective against pesticide-resistant pests.

Some disadvantages of this method include;

- It is generally slow especially in the beginning.
- Low levels of infestation, with occasional outbreaks, remain a feature of systems under biological control.
- Any use of chemicals around the biocontrol agent may greatly affect the potency of this control method.
- It cannot be used where the biocontrol agent would threaten commercial populations of the target species that may exist nearby.
- Lastly, biocontrol agents are not available for all target IAP species, therefore, it is not always applicable.

Biological methods have proven to be very effective in the elimination of pests that cause soil-borne diseases.

8.4 CHEMICAL CONTROL

This is the control of pests using pesticides. Pesticides are chemicals used to prevent, repel or destroy pests. They combat pests and diseases occurring on crops, livestock and our possession. They are applied in the soil as pre- and post-plant applications. Mostly given to high-value cash crops. They are applied as soil fumigation, soil drenching, and seed treatment. They are classified according to their usefulness. Acaricides are used to control ticks and mites. Insecticides are used to control insects. Fungicides are used to control fungal diseases in plants. Herbicides are used to control weeds. Nematicides are used to control nematodes. Lastly, rodenticides are used to control rats.

They have three different modes of action;

- Contact poison: the action of these chemicals follows the action between the contact of the body of the insect insecticide e.g. Carbaryl.
- Systemic poison: these chemicals move within the plant tissue and toxicity to sucking insects feeding on plant sap e.g. Carbofuran
- Fumigants: this acts by its vapor action e.g. Methyl bromide.

Botanical compounds e.g. Pyrethroids are produced by the ground flowers of the daisy. Synthetic organic compounds such as organochlorines contain C, H and O. Organophosphorus is made up of organic molecules containing phosphorus. Carbonates are structurally esters of unstable carbonic acid.

Microbial compounds are commercially produced insecticides from the natural pathogens of insects. Synthetic pyrethroids are synthesized from petroleum-based chemicals (Agtmaal, 2015).There are numerous ways for the application of pesticides, but the knapsack sprayer is the most popular spray equipment.

Some methods of soil treatment are soil drenching, furrow application, spot application, strip application, pits application, planting site application and broadcasting of dust, powder orgranules.

Some advantages of chemical spray control are;

- It is cost-effective.
- It is not time-consuming and allows flexibility.
- There areQuality and quantity for the price of the produce.
- It prevents any problems with pests.
- It does not damage the environment. It is environment-friendly.

Some disadvantages of chemical spray control are;

- It causes a reduction of beneficial species. Nontarget organisms, including predators and parasites of pests.
- It leaves residues in food for humans and feeds for livestock.
- It causes groundwater contamination by leached chemicals.
- It has poisoning hazards and other health effects.
- The drift of sprays and vapor during application can cause severe damage and residue problems in crops.

Fungicides are important in preventing and eradicating plant diseases. Some advantages are, they are more specific inhibitors of the biosynthetic process, they cover foliage and inflorescence more effectively, they are translocated thus used in lower concentrations and they provide control when pathogen already established. Movements of fungicides in plants are very unique and specific. There is apoplastic translocation which is upward movement by the xylem. Symplastic translocation represents a downward movement by phloem. Ambimobile represents up and down both side movement. Toxophere talks about many systemic fungicides are metabolized or degraded inside plants (Agtmaal, 2015).

Weed control using chemicals has a few limitations. The chemical method has led to many problems such as weed flora shift, the insurgence of herbicide-resistance weeds. The spectrum of weeds killed by herbicides is low due to the growing resistance. Herbicides, therefore,cannot be the sole and foolproof strategy in weed control.Chemical control of pests has shown its ups and downs, but one still has to consider all the available options before choosing a method of disease control.

8.5 HOST PLANT RESISTANCE

Host plant resistance to insects is the phenomenon and interrelationship between host plants and is affected by environmental factors. It is governed by a single gene or a small number of related genes, which encode proteins capable of altering the outcome of an otherwise compatible plant-pathogen interaction. The biochemical mechanisms responsible for the induction of specific resistance in plant-pathogen interactions are generally poorly understood but are likely to vary with both the type of specific resistance and the plant-pathogen system involved. The three most common mechanisms underlying specific resistance appear to be race-specific elicitors induced only to a particular race of pathogen present in all cultivars of the host plant. Host-selective toxins are due to the production of compounds that are toxic to plants. Race-specific suppressors are due to pathogen production of race-specific suppressors that inhibit a non-specific resistance response (Puri, 2016).

There are many important reasons to focus on host plant resistance. It is important to develop techniques to screen for resistance to insect pests under optimum levels of infestation and under similar environmental conditions. Techniques to screen for resistance to insects have been developed against some insect species. However, there is a need to develop systems for insect rearing, refine the resistance screening techniques and establish long term programs to screen and breed for resistance to several insect species that are important crop pests.

Advantages of host plant resistance are;

- Resistant variety is good and is not harmful to the environment.
- Resistance is persistent until it is broken by something specific.
- It does not cause any pollution and is harmonious with the environment.
- It is host-specific.
- It is compatible with other methods e.g. Resistance variety – early/ late sowing.

- The number of insecticide applications reduces.
- Keeps the pest population at a low level.

Some disadvantages are;

- It takes a longer time to achieve.
- There is a genetic limitation: it is not able to cross with commercial species.
- It cannot solve the current rising pest problems.

The mechanisms of resistance are; Antixenosis which is the response of the insect to the characteristics of the host plant which makes it unattractive to the insect for feeding, oviposition or shelter. Antibiosis is the adverse effect on the insects' life history which results when the insect uses a resistant variety of the host plant or species for food. Tolerance is the ability of the host plant to withstand the insect attack and grow satisfactorily despite the attack, by rapidly repairing the damage or by quick development of new tillers and roots (Puri, 2016). Adaptations of resistance are either morphological, anatomical or biochemical traits. Morphological factors include surface waxes, color, shape, size, trichomes on plant surface, and hardness of plant tissues, etc.

Chapter 9 Association Between Soil-Borne Pathogens and Global Climate Change

CONTENTS

9.1 INTRODUCTION

Climate is the average condition of weather over thirty to thirty-five years. When the weather condition for several years is noted down, then the average condition is calculated, then the results give the climate. Climate can be classified according to climatic zones. There are five major climatic zones all over the world. They include:

- Polar climate zone which is ever cold and dry all over the year. They are areas at the North Pole, South Pole and the interior of Greenland.
- Temperate regions- these are the regions whose winters are cold, and summers are mild. This covers Northern America most parts, Asia to the north and Europe.
- Arid zones- these are the ones that are ever hot and dry throughout the year. In these regions, the climate is hot and dry and only supports shrubs and short woody trees. This climate covers all the deserts in the world.
- Tropical regions- in these regions, the climate is hot and wet. It, therefore, means that these regions experience sunlight for about 12 hours, meaning that they support many numbers of tree species. It covers several parts which lie between the tropic of Capricorn and the tropic of Cancer.
- Mild Mediterranean- these are the regions thatexperience mild winters and hot and dry summers. Mediterranean Sea surroundings experience this kind of climate.

For the climate to be determined, several patterns of weather should be observed. Now when the global and environmental pattern changes, this is referred to as climatic change. It takes several decades for climatic change to occur and once change occurs, it may last for a bit longer period like millions of years (Rana & Randhawa, 2015).

9.2 CAUSES OF CLIMATIC CHANGE

There are several factors which cause climate change. Below are some of the causes of climatic changes:

9.2.1 Human Activities

Human activities are one of the main factors influencing climatic changes all over the world. Human beings carry out different activities. Deforestation is one of the events that a man does that can affect climate change in a specific region. Deforestation refers to the cutting down of trees. It is well known that trees attract rainfall. In most areas where there are many trees, either natural or man planted, there are adequate rainfalls. This means that trees have a very paramount purpose in ensuring the right climatic conditions. When the trees are cut down, it means that the rain will decrease. A decrease in the amount of rainfall automatically leads to climatic change. Most likely, the climate will change to be arid. Afforestation is another human activity that can influence climate change. Afforestation is the aspect of planting new trees in the places where there were no existing trees. When new plants are planted, they lead to increased rainfall which in return influences climate change (Rana & Randhawa, 2015).

9.2.2 Greenhouse Effect

The greenhouse effect is the process by which radiations warm the earth from the sun. The greenhouse effect is caused by greenhouse gases which damages the ozone layer. These gases are referred to as chlorofluorocarbons. There are other Gases that are produced by the greenhouse such as methane, carbon dioxide, nitrous oxide, and ozone. Ozone forms ozone layer which prevents the violent and harmful rays from the sun from reaching the earth's surface. These rays can kill human beings, animals and even plants. The chlorofluorocarbons produced by the greenhouse are harmful to the ozone layer, and therefore it damages it by creating some spaces. Damage of the ozone layer then means that the harmful sunrays will penetrate the earth's surface. Due to the irregularity of the earth's surface, these rays may not be able to be reflected. Therefore, this causes global warming thus affecting climatic changes (Rana & Randhawa, 2015).

9.2.3 Ocean currents

Also, ocean currents can, in a way, cause climate change in some regions. Oceans and the other tropical water play a very paramount role in absorbing the heat from the sun. They consume almost sixty percent of the heat from the sun. Therefore, oceans and the tropical water masses regulate the earth's warm through its breezes. In the oceans are the ocean currents which are initiated by the winds over the passing over the seas. Some of these winds include the monsoon wind. Since the ocean water has got some heat from the sun, the currents carry with then the warmth which may affect climate change to regions surrounding the oceans depending on the direction of the winds. The ocean currents at other times are cold, and therefore, this regulates the temperatures of the equatorial and tropical regions, affecting the climate ultimately (Rana & Randhawa, 2015).

9.2.4 Vegetation Cover

In most areas, the vegetation covering the land correlates with the type of climate to be found in that region. This means that vegetation cover can cause climate change by changing the nature of vegetation in that region.

9.3 IMPACT OF CLIMATE CHANGE ON THE SOIL FUNCTION

Change in climate in a region affects the soil structure either positively or negatively in numerous ways. The climate of a given region plays a very crucial aspect in determining the soil structure and fertility of the soil in that region. For instance, climate change affects soil moisture. When there is deforestation which may lead to a hot and dry climate, the moisture content in the soil drops dramatically. This alters the soil structure of the soil since some of the living organisms which ensure the build-up of soil structure cannot be effective in a place without adequate moisture content.

Besides, soil living organisms such as earthworms help in enhancing soil structure. These organisms survive and work efficiently depending on the nature of the climate and other factors in the region they dwell. Climate is the most affecting factor. When it is wet, organisms such as earthworms do better and thus perform their activities efficiently, thus improving the soil structure and fertility. On the other hand, when the climate changes to hot and dry due to some effects, this affects the earthworms, thus slowing their activities since some will burrow themselves down to take refuge. Others can even die, thus decreasing their numbers. Reducing their numbers means a slower rate of building-up the soil structure (Rana & Randhawa, 2015).

Again, climate change alters the recycling of nutrients and modification of soil biology and chemistry. Some activities, such as nutrient recycling is carried out by soil organisms such as earthworms. The earthworms, therefore, depends on the climate to carry out their activities. Changing climate thus influences these earthworm's actions. When the soil is wet, they carry out the activities smoothly, unlike when the climate is hot and dry, thus affecting the soil. High rains also change the structure of the soil by water-logging. When high rainfalls occur, the water enters and closes the burrows dug by the earthworms which ensure sufficient air in the soil. Logging of these warrens by the water means that the soil will be dense and with no adequate air, thus affecting the soil supplies to the plants.

Change in climate affects the yield produced in a particular region. As it is known, in places where there are adequate rainfalls and cold and wet climate, most crops can do well and produce high yields than in areas with inadequate rains. This is caused by the fertility of the soil depending on the climate of the place. The earthworms deposit their cast on the soil surface thus improving soil fertility. With wetness, the climate favors more worms which improve soil fertility than in dry areas (Rana & Randhawa, 2015).

9.4 IMPACT OF CLIMATE CHANGE ON DIFFERENT SOIL PARAMETERS

Figure 9.1: Changing global climate patterns will affect soil fertility and productivity.

Source: https://www.organicconsumers.org/news/how-worlds-most-fertile-soil-can-help-reverse-climate-change-0

9.4.1 Physical Parameters

Soil water: Human activities often result in alterations in the earth's environment, which could sometimes cause severe outcomes in soil water composition. Fluctuations in the atmospheric gas composition, partly because of CO_2 and other greenhouse emissions, may cause an increase in temperature levels with high altitudinal and sequential variability. These factors may trigger a serious reorganization of atmospheric precipitation trends, escalating aridity in some regions due to lesser rainfall. Moreover, decline in atmospheric precipitation caused by manmade actions may result in a reduction in water infiltration, as well as storage within the soil. This is inclusive ofplants' water supply which shall be affected due to surface runoff, especially in hilly terrains with rolling surfaces.

Also, water erosion hazard is probable due to unsustainable water levels which some areas may experience as a result of climate change. Even so, if the soil has a good plant cover then evaporation losses would be controlled, as compared to a barren land which shall lose most of its moisture to the environment. Other factors that determine a soil's capacity to withhold moisture are its type, density, biomass production, and root characteristics. Global warming affects the water cycle by altering regional precipitation levels, including temporal variability, plus also impacts on water movements and soil humidity dynamics. Infield studies, increasing average yearly temperatures and declining precipitation levels during summer season have been observed in some regions. In tilled lands, available soil water levels were generally higher, while trends in soil humidity dynamics were largely balanced due to a decent greenery cover and favorable land conditions (West et al, 2015).

Soil temperature: Soil temperature (ST), is a value that represents the thermal status of the ground within a given location. It's a marker of climate disturbances brought about by human activities. In fact, knowledge of ST factors through the soil condition is regarded as an effective way of measuring the extent of climate change dynamics in a particular region accurately. Additionally, freezing and melting processes of the soil that occurs all year round play a crucial role in shaping the composition of land and atmospheric interactions. From scientific observations, conclusions may be derived that heat exchanges during soil freezing are thrice more rampant in regions that experience seasonal climate fluctuations, compared to those without these phase changes that are mostly found around the equator.

Observations on northern regions also show the considerable diurnal variances in soil temperatures, caused a diurnal sequence of unfrozen moisture content at the soil surface. Further studies show that the freeze-thaw cycle may actually be a buffer for seasonal alterations in soil, including near-surface temperatures and reinforces the energy interchange between the soil and environment. An increase in mean annual atmospheric temperatures, caused by global change, consequently results in widespread permafrost degradation in northern territories, while other areas may experience increased rainfall due to higher vapor levels in the air. Lately, there have been several reports of soil temperature variations linked to climate change in different areas of the world, like Alaska, Turkey, Canada, Russia, China and Siberia. Temperature is a direct indicator of climate change, including the freeze-thaw cycles of different regions (West et al, 2015).

Soil structure and texture differentiation: Soil structure controls its physical health and hence plays a vital role in determining productivity, including offering on-site and off-site ecological support. Potential alterations to the various factors of soil structure, form, firmness and resilience could result from environmental changes. Studies reveal the potential dangers of structural soil degradation, coming from both indirect and direct sources of climate change. These are particularly intense for hard setting soils found in sub-tropical and temperate areas. The subsequent soil structural degradation may lead to limited rainfall infiltration, improved soil erosion and minimized water usage efficiency with inferences on productivity, as well as ecosystem. A few of the negative impacts in soil structure may already be evident in vegetation disturbances and hydrology. Even so, with adequate soil composition management, agroecosystems can be adopted to accommodate future climate changes.

Presuming constant additions of carbon into soils from plants, soil-climate experts estimate that projected changes in temperature, humidity and evaporation will trigger considerable upsurges in organic material turnover and greater losses of carbon-dioxide in mineral and natural soils across different terrains.This often results in constructive feedback between carbon-dioxide discharges from soils and additional temperature increases. These wide predictions have been partly supported by the variations in existing soil carbon stocks, which have been measured by scientists. Losses in soil carbon will further impact on other soil functions. hHighest losses, comparative to prevailing soil carbon content, exist where temperature increment rates are the greatest. As such, this may result in weaker soil structure, firmness, topsoil moisture retention capacity and nutrient availability.

9.4.2 Biological Parameters

Soil organic matter: Soil organic material stocks are often mentioned as an option for limiting greenhouse gas releases. Nevertheless, growing carbon stocks present in agricultural soils may become difficult to manage, including the capacity of soils to retain carbon as climate change factors continue to fluctuate. This could further be due to interacting features, such as those varying spatially, and contributing to organic material inputs and disintegration rates. In a study involving two models, which are sustainable grazing systems and holistic-farm system models, where the potential impact of environmental changes on soil organic matter was assessed. Results showed that expected rainfall, the attribute with the greatest variance between climate settings, was the main source of ambiguity in organic content response. Otherwise, dry climate conditions caused lowered material composition compared to other scenarios.

These two models presented more or less similar trends, though the latter model in overall gave greater percent changes for soil organic matter over the testing period and a greater range of responses as a result of changes in land site features, especially clay content. Going into the future, wet climatic conditions are expected to cause improved production, which shall support upsurges in organic material stocks. Conversely, soil organic matter will accumulate at much slower rates, or continue remaining constant under dry projections, plus lowered production further means that this may be followed with decreased average organic material stocking rates among farmers. Additionally, climate changes would likely be associated with irregular average root mass and exudation, mycorrhizal colonization, including the activity of various rhizosphere and soil microbial communities (Rana & Randhawa, 2015).

Soil fauna and soil flora: Generally, countries found in northern territories, like Russia and Canada would be the most predisposed to soil flora and fauna disruptions resulting from climate changes. As plant pathologists predict, the main threat to microbiology will result from heightened frequency and scale of abiotic stress factors like frost, drought, flooding, heat, and temperature fluctuations. Plus, there would be increased spreading of pests and pathogens across different areas, they shall be attracted to expanding ecological conditions favorable to them that climatic changes will provide.

There are several examples of rapid spreading of thermophilic plant pathogens that have been recorded in Russia, some of them being Erwinia amylovora, Acidovoraxsp., Dickeyasp., phytoplasmas and Ralstonia solanacearum along with various plant virus species. Furthermore, thermophilic fungi such as Fusarium and Vericilliumhave become more predominant soil-borne pathogens, spreading to new areas that they previously didn't inhabit. Basically, there shall be both winners and losers in climate-change dynamics. Some parts of the globe will develop conducive habitats for particular soil-based species, for instance, as some colder zones become more tropical, species that thrive in warmer climates will possibly migrate to those areas (Rana & Randhawa, 2015).

Nevertheless, climate change is expected to reduce the size of various soil creatures and plant populations, including some microbes which provide vital nutrition. From micro-organisms to higher forms of life occupying the soil, around 45 percent of the species sampled by researchers have been found to be growing smaller over successive generations, largely part of climate change. Carbon pollution, in particular, has contributed to around 1°C increment in global temperature levels.

9.4.3 Soil Chemical Parameters

Chemical processes in soils: Soils form part of the worldwide carbon and nitrogen exchange cycles. All carbon-based gases including, methane CH_4 and carbon dioxide CO_2, as well as nitrogen-based emission nitrous oxide N_2O, that are vital greenhouse gases. Therefore, as their levels shift in the environment, there would be corresponding variations in the soil as well. The main challenge for soil researchers today is figuring out precisely how climate alterations will affect soils since some probable effects cross each other out. Case in point, organic material is helpful in soils, and scientists presume increased CO_2 levels in the environment could boost plant growth, leading to potential upsurge in soil organic matter as well.

Nevertheless, greater temperature also translates to heightened rates of organic material decomposition through soil microbes. In case the microorganisms disintegrate organic matter much faster than it's replaced, overall soil organic matter portions will decline as well. From these explanations, it's evident that there are opposing effects of climate change on soil chemical processes, some are beneficial while others are not. According to scholars, worldwide temperatures are projected to rise between 1.1-6.4 °C in the 21st century, plus precipitation patterns shall be altered as well.

Since soils are delicately connected to the climatic system through the gas exchange cycles, weather alteration will definitely impact on their chemical processes and properties (Rana & Randhawa, 2015).

Other studies show at least various types of soils will become net contributors of atmospheric Carbon, reducing soil organic matter content considerably. Soil erosion triggered by erratic wind and rainfall patterns will also likely increase over the next coming years.

Acidification, salinization, sodicity problem in soil: Variable and shifting climate will affect soil properties, such as pH which is a top variable that affects every other factor in the ecosystem. There's minimal knowledge regarding the specific outcomes of temperature fluctuations and rainfall trends on soil properties, but increased acidification is believed to be among them. Intensified leaching of primary cations because of increased rainfall shall diffuse alkalinity, coming from soil exchange complexes into both ground and surface waters, leading to acidified soils. Furthermore, altered water conditions in soils consisting of oxidizable sulfide matter may cause heightened soil acidification, or acid sulfate soils, in case these soil layers get exposed to air. A greater understanding of the outcomes of shifting and variable climatic conditions on soil pH, including other soil features is required to inform impactful decisions on inhibition of terrestrial ecosystem losses. Mineral changes in the soil may be instigated by climate change, leading to loss of salts including nutrient cations on the ground. Increased evapotranspiration or decreased rainfall levels are some of the main causes of soil salinization. Areas that may have previously been under perennial moist conditions may become subject to periodic drying, even gradual desiccation of important compounds like goethite and hematite which shall be affected by higher temperatures or intense drying conditions (Rana & Randhawa, 2015). Likewise, sodicity is another major cause of soil degradation in many places. It's caused by increased concentrations of sodium that generally get fused to clay particles. Consequently, clay particles found in the soil will lose their capacity to fuse together when moist. This causes unstable soils thatcould erode or turn to be impermeable for water and crop roots to hold on to. Soil fertility and nutrient acquisition: Climate change variables, such as precipitation, increased temperature and atmospheric carbon-dioxide concentrations will alter soil fertility patterns globally. CO_2 is a plant compound, and its environmental enrichment shall potentially lead to enhanced crop productivity. As some researchers have noted, agriculture might be among the few bright spots or 'silver lining' in climate change, however, increased global temperatures and disrupted precipitation patterns are also anticipated to accompany these escalated CO_2 levels. These factors may diminish or negate any crop production surges or even reduce production below existing levels.

Interactive influences of environmental carbon dioxide levels, temperature and soil humidity on crop growth, maturity and yield have well been documented. Climate change factors influencing these particular variables will affect crop water use efficiency (WUE), including physiological and uptake abilities of soil-immobile nutrients. Increased atmospheric CO_2 will further upsurge leaf and plant canopy photosynthesis, but with trivial alterations in dark respiration. Additionally, carbon dioxide will considerably intensify biomass without any marked deviations in dry material partitioning, further reducing transpiration levels of many crops and enhancing food production. Even so, spatiotemporal disparities in these attributes shall greatly impact on agronomic performance, including crop water usage in a site-specific way (Rana & Randhawa, 2015).

The nutrient acquisition may as well be affected, considering that it's closely tied to general biomass and highly affected by root surface structure. There's growing evidence to suggest that changing global climate factors will affect the physiological capacity of nutrient usage within the plant, while further altering the concentration of soil nutrients and their transfer through the soil right across root membranes.

9.5 MICROBIAL CONTRIBUTIONS TO GREENHOUSE GAS EMISSIONS

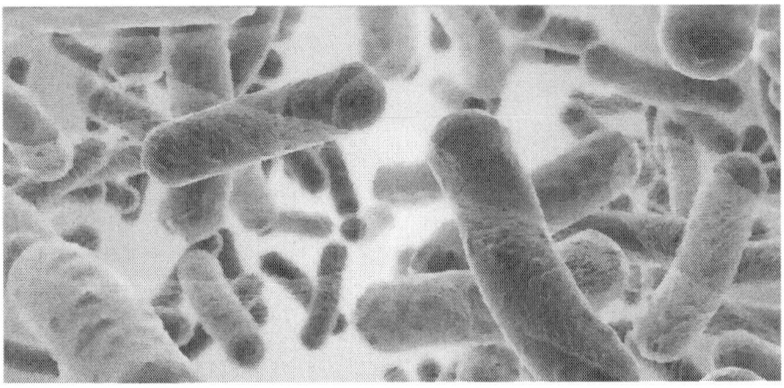

Figure 9.2: Microbes act as catalysts by neutralizing greenhouse gases in the soil.

Source: https://study.com/academy/lesson/ecological-role-of-microorganisms.html

Soil microbes mediate nearly all processes that lead to the generation or utilization of green-house gases present in soils. For example, inadequately drained soils, methanotrophs may intake CH_4 plus soils further serve as the drain for CH_4. Alternatively, in ill drained soils, methanogens have the potential to produce CH_4 through utilizing by-products resulting from anaerobic degradation of soil organic material, plus these soils may become a basis for CH_4.

Generally, the production or utility of N_2O in cultivated soils is determined by two forms of soil microorganisms, which are denitrifiers and nitrifiers, whereas multiple heterotrophic microorganisms are engaged in organic matter breakdown and CO2 emissions. Without any doubt, detailed knowledge about soil microorganisms participating in these processes would be helpful in designing suitable mitigation measures for decreasing these emissions (Tanmoy et al, 2016).

It might not be simple to manage soil microbes directly; though, microbial processes causing greenhouse gas emissions in soil could be managed remotely through appropriate soil management including agronomic practices. Case in point, zero-tillage may reduce the amount of organic material decomposition and carbon dioxide emissions; additionallydenitrifiers might be managed secondarily or through the manipulation of soil water aspect, even manipulation of nitrate supply through the utilization of nitrification inhibitors.

Mineral application to farming soils can be controlled to maximize efficient use by plants while decreasing N2O emissions. Some best management practices that can be applied include; Right rate,place, time and source. Of all the major greenhouse gases, which are CO_2, N_2O and CH_4, carbon dioxide is the principal gas that contributes to the overall soil greenhouse gas (GHG) profile, closely followed by N_2O. Within most agricultural soils, apart from flooded rice fields, emissions from CH_4 towards the overall GHG profile areimmaterial (Tanmoy et al, 2016).

For instance, the relative contribution of 3 gases to overall soil GHG within whole maize growing season in Northern America and Canada is estimated to be roughly 90% carbon dioxide, and 10% N_2O. Nevertheless, there's still need to take heed that nearly all of the CO_2 produced may be newly fixed carbon by photosynthesis occurring in corn. The gases could be resulting from plant residue decay and root respiration.

Hence, for a thorough assessment of the overall global warming effect of this cropping structure, alterations in soil carbon must be examined over multiple years. While photosynthetic bacteria may capture carbondioxide from the environment directly, still their contribution to carbon balance in cultivated lands is negligible in comparison to photo-synthetic carbon fixed by plants (Tanmoy et al, 2016).

9.5.1 Greenhouse Emissions Management

Biofuels derived from decomposing plants, algae and fungi may be useful in avoiding the need for environmental destruction, either as cellulosic or algal biomass, which are significant elements in biofuel production. The method carries various advantages, including the achievement of a cleaner and eco-friendly environment. It recycles carbon-dioxide trapped in plants, thereby leading to reduced greenhouse gas emissions over the long run. Also, it uses solar power to change the CO_2 into a fluid fuel. The fuel is a major improvement compared to other modern alternatives thatcontribute to the carbon footprint. Therefore, while some microbes increase the numberof greenhouse fumes released, others may actually be used to suppress them.

9.5.2 EFFECTS OF ELEVATED CO_2 LEVELS ON SOIL-BORNE PATHOGENS

Soil carbon might not be as steady as previously assumed, researchers report, further arguing that soil microbes put more direct regulation on carbon accumulation than international climate models actually show. This study, researchers claim, delivers insight into the systems determining longstanding soil carbon storage, understanding that can be applied to develop more accurate climate model representations touching on the international carbon cycle. Increased crop growth caused by heightened atmospheric carbon dioxide, has been linked with higher CO_2 release rates from soils. In case the rising carbon dioxide improves soil carbon storage levels at all, then the effect shall be insignificant,.

Basically, carbon dioxide is effectual in prolonging the lifespan of plants by impeding bacterial proliferation. The general effect of carbon-dioxide is increasing both the lag stage and the generation period of microorganisms responsible for decay; nonetheless, the specific method for this bacteriostatic effect is still unknown. Displacement of oxygen, including intracellular acidification are two possible methods that have proposed, but researchers are still skeptical about them. Swift cellular penetration and change of cell permeability features have further been discussed, though their connection to the general mechanism is not yet clear. Some scientists have suggested that CO_2 may first be assimilated into a liquid form of the treated cell to develop carbonic acid (H_2C0_3), plus investigations by scholars further confirm this step, while also indicating a possible direct benefit of carbonic acid in retarding harmful bacterial activity (Tanmoy et al, 2016).

Most lately, a metabolic process has been scrutinized by researchers whereby CO_2 in the soil pathogen cells has damaging effects on different enzymatic and biochemical activities. The joint results of these metabolic interferences have been thought to contribute towards stressing the system, resulting in a reduction of the harmful microbes' growth rate. Even more, the extent to which CO_2 is effective mostly increases with its concentration levels, though very high amounts raise the chances of creating conditions where pathogenic microbes like Clostridium botulinum could survive. It's believed that such hazards, however, can be lessened with proper temperature regulation of soil conditions where the pathogens live.

9.5.3Mechanism of CO_2 Action

One of the ways CO_2works is by displacement of oxygen, which starves pathogen of breathing air and suffocates them. It further has the capacity to displace some or all oxygen present for microbial metabolism, thereby slowing down their growth comparatively. Research also shows that anaerobic bacteria areinhibited by CO_2 atmospheres, where under the right conditions bacterial growth environments can greatly impede.

Although decreasing available oxygen could have certain effects on bacterial growth, it seems not to be the greatest limiting factor for microbial propagation. Changes in soil pH soil seem to have a more significant effect on managing this factor. Otherwise, for carbon dioxide, it's usually dissolved first into liquid form of the tissue, before being absorbed back as carbonic-acid in its undissociated form. This process for movement of CO_2 into an organism's cell may help explain researchers' observations of rapid pH decrease in the tissue following carbon dioxide entry. Internal acidification has further been identified as the main reason behind bacteriostatic effect (Tanmoy et al, 2016).

9.6 EFFECTS OF TEMPERATURE ON SOIL-BORNE PATHOGENS

Typically, the temperature increase wills fast track enzyme activity inside the pathogen. But if the heat is too high, then enzyme function will weaken and the protein or enzyme shall denature. Alternatively, lowering temperature shall drop enzyme activity and at freezing levels enzyme action may cease completely. Also, repeated phases of freezing and melting can destroy catalyst proteins. Additionally, freezing makes water to swell and also develop ice crystals, thus pathogen cells may begin to rupture. Each microbial species has certain growth temperature needs, which are mostly determined by specific temperature requirements its enzymes must have to function optimally. Basically, all organisms have their own least possible growth temperature, ideal growth temperature, and maximum temperature level (Tanmoy et al, 2016).

9.6.1 Pathogen Classification According to Temperature

Microorganisms can roughly be categorized depending on the temperature range in which they can mature. The growth capacities are highest at the optimal growth temperatures for that particular organism, whereas in the lowest and maximum temperature ranges they might only survive or reproduce at negligible rates (Tanmoy et al, 2016).

Mesophiles are organisms thathave adapted to modest temperatures, with optimum growth temperatures that range from between room 20 °C to around 45 °C. As expected from the average temperatures of a person, which is about 37 °C, most human-based microbes and pathogens are ranked under the mesophiles group. Some common ones areLactobacillus spp., Salmonella spp. and E. coli. Microbes called psychrotrophic, are also referred to as psychrotolerants, these ones prefer cooler environments, with high temperatures of 25 °C up to freezing conditions of about 4 °C. These germs are found in various natural environments mostly in temperate climates. On the other hand, psychrophiles or cold-loving microbes are designed for growth in conditions below 0 °C though their optimum growth temperature is roughly 15 °C, usually don't survive at conditions above 20 °C.

Organisms that mature at optimum heat levels of 50 °C to the maximum of 80 °C may also be called thermophiles, or "heat-loving" microorganisms. They don't reproduce at room temperatures. These thermophiles are extensively distributed around hot springs, manmade ecosystems and geothermal soils like compost piles where they normally breakdown household scraps and vegetal matter. Some examples of thermophiles are Geobacillus spp. and Thermus aquaticus. Much higher on the temperature scale there are hyperthermophiles, these are marked by growth rates of between 80 °C up to a top-level of 110 °C, while in extreme cases there are those which can withstand temperatures of above 121 °C (Tanmoy et al, 2016).

9.6.2 Temperature Change Effects

At very low temperatures, organism cells can be affected in various ways. First, membranes will lose their flexibility, plus they are destroyed by ice crystal development as well. Molecular reactions and dispersal slow considerably, while proteins get too dense to catalyze chemical reactions and may go through denaturation. On the opposite scale of the spectrum, excess heat also denatures proteins including nucleic acids. Improved fluidity affects metabolic reactions in membranes. Some practical steps that can mimic the damaging effects of high temperature on microbes include; steam sterilization, pasteurization, and burning of inoculating loops.

9.7 EFFECTS OF DROUGHT STRESS ON SOIL-BORNE PATHOGENS

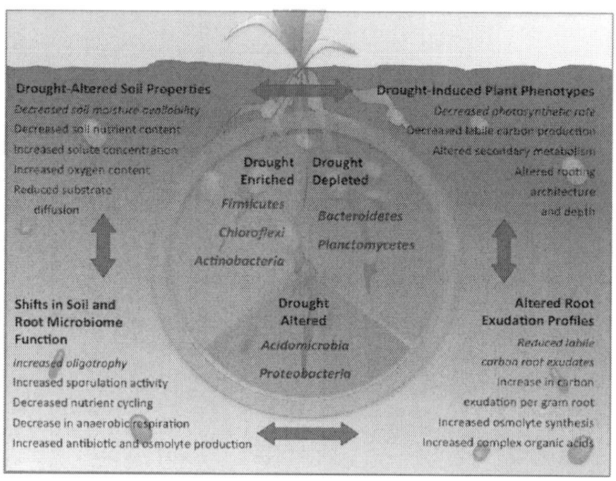

Figure 9.3: Drought stress can alter the behavior and activity levels of soil microbes.

Source: https://www.frontiersin.org/articles/10.3389/fpls.2017.02223/full

Drought is recognized for bearing strong effects on soil composition, including the activity of pathogen communities, which can be defined based on the drought history and its overall duration, though the characterization and estimation of these effects still remain to be a test. This is mainly because, soil pathogen communities that have formerly been subjected to drought, can less alter their nature in response to successive drought events, because of the range of drought-resistant species.

Generally, soils that have a long drought history experience greater overall microbial alpha diversity, presumably due to adaptation of the microscopic community to drought environments. Nevertheless, some pathogenic taxa like Chloroflexi and saprotrophic yeasts actually showed diminished abundance in pre-droughted soils. These studies reveal that soil communities may not necessarily have the capacity to preserve similar functions, especially during extreme or recurrent future droughts, whereby microbial functions are defined by the soil community composition. Drought is still likely to continue affecting community structure, even in soils thatmay be acclimated to it, basically tending to grow the population of fungi while reducing the ratio and diversity of microbes (West et al, 2015).

9.7.1 Case Study

Drought stress has been ranked among the greatest obstacles in farming. Helpful microbes found in the root structure of some crops, may not be able to withstand the drought effects on growing plants. Arid conditions primarily rise the abundance, including activity levels of monoderm Actinobacteria within the soil around the root system and plant tissue.Scientists from the University of California, Berkeley and other research partners studied the pathogen community as regards sorghum. They scrutinized how the crop's microbiome and root metabolome altered across that entire period. They discovered that arid conditions primarily increase the profusion and reactions of monoderm Actinobacteria. In drought conditions, the monoderm microbes increase transcription capacity for genes associated with metabolite transport. It was concluded that under drought stress sorghum releases metabolites that choose appropriate microbes in the adjacent root area, these, in turn, may help the crop to better resist drought stress. This finding may hold intuitions into how crops like sorghum control or influences the soil microbes, which consequently promote drought resilience in plants (West et al, 2015).

9.7.2 Developing Trends

- The outcome of drought on soil organisms will vary across different taxonomic groups, though they remain constant across all seasons.

- Furthermore, the anthropogenic global change affects the activity and structural composition of soil-borne communities, which are liable for vital ecosystem systems and functions. Some of the most prevalent global change factors are drought and mineral enrichment. Nonetheless, the reactions of soil micro-organisms to interacting universal change factors arestill largely unknown.
- Experts agree that drought has unfavorable effects on soil-borne invertebrates feeding activity, including a simplified nematode community system, while surprisingly in some cases soil microbial action and biomass remained completely unaffected.
- In general, microbial biomass increased as a reaction to fertilization, while invertebrate feeding action noticeably declined. Markedly, these effects were regular across seasons. These dissimilar responses further show that soil biota variesimmensely in their susceptibility to universal environment change drivers. Therefore, important ecosystem procedures like decomposition and mineral cycling that rely on soil microorganisms could be interrupted with drought conditions.

9.8 MECHANISMS TO SOLVE CLIMATE CHANGE

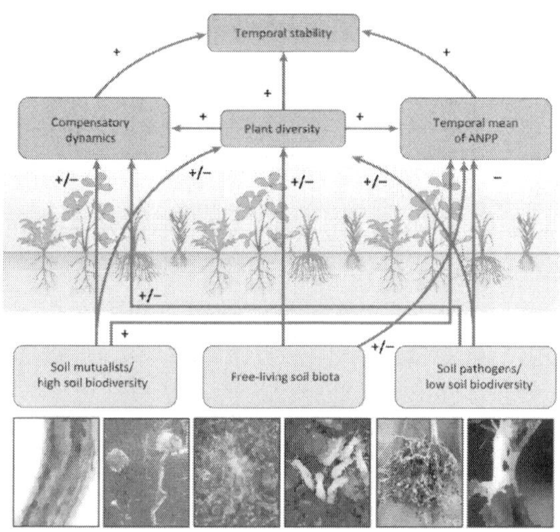

Figure 9.4: Soil microorganisms help to stabilize the ecosystem by managing the release of greenhouse gases.

Source: https://www.cell.com/trends/plant-science/fulltext/S1360-1385(18)30212-7

9.8.1 Microbial Communities and Carbon Cycle

Global factors like greenhouse gas emissions, including human and natural, caused factors like coal combustion and other fossils, plant matter decay and biomass burning all affect microbial communities in the soil. In particular, global warming and climate change have been shown to damage a number of biological processes. However, some microorganisms also have a potential role in mitigating emissions. They are especially applied in carbon and greenhouse fumes treatment or reduction from nutrient recycling. Carbon gas emissions have increased drastically in recent years because of human activity, as well as natural occurrences like volcanic eruptions (West et al, 2015).

As one of the main greenhouse gases, carbon dioxide is massively produced during fossil fuel usage across different sectors such as transportation, construction, heating, cooling and production of heavy-duty goods. It may also be released from organic processes like decomposing plant material, respiration and microscopic breakdown of organic matter. The gas accumulates in the environment over time, causing toxicity to consequently increase and pollute the ecosystem.

Carbon cycle effects: Carbon found in the globe's atmosphere occurs in two key forms, which are methane and carbon dioxide. CO_2 oftentimes is dissolved directly from the environment to land and water bodies. Furthermore, it may dissolve in precipitation when raindrops drop through the atmosphere. Once dissolved in water, CO_2 reacts with moisture molecules to form carbonic acid that contributes to the ecosystem's acidity. Microbes form part of the larger cyclic chain of carbon which happens on a global level. The actions of helpful germs help to remove carbon from lifeless sources and turn the carbon to a form that's accessible for living organisms. Before carbon-dioxide can be assimilated into living-organisms, it should first be changed into a usable organic format. The transformative procedure by which CO_2 is retrieved from the atmospheric system and added into organic materials is known as carbon fixation.

Among the best-known cases of carbon-fixation is photosynthesis, which is a procedure whereby energy gotten from sunlight is utilized to develop organic compounds. In particular, photosynthetic algae are vital microorganisms in this respect, including the chemoautotrophs. Mostly, bacteria and other pathogens are capable of converting carbon dioxide in the soil to form sugars which are crucial for cell building. While some organic carbon may be returned back to the environment as carbon-dioxide form during respiration, a good portion of the environmental carbon may be recycled from one organism to another throughout the food chain (West et al, 2015).

In case an organism perishes, it's decomposed by microbes and its carbon residue released back into the ecosystem of the soil. This gas dissolves in water and during the same time algae, bacteria and plants transform it into organic carbon that's useful for themselves or other creatures in the ecosystem.

Carbon transfers in a cyclic way between creatures, from its producers to consumers and all over again. Cells and tissues are typically broken down by microbes and CO2 arefreed back into water bodies or the atmosphere, where other organisms reuse it once more. Both aerobic and anaerobic processes are involved in the carbon cycle across different ecosystems (West et al, 2015).

9.8.2 Microbial Communities and Methane Cycle

Methane is a crucial anthropogenic greenhouse fume, typically produced and utilized in soils by microbes responding to ecological conditions. Estimates reveal that soil consumption covers for about 5–15% of methane, which is discharged from the environment annually. Researchers have found sufficient evidence for connections between changes in the methane-cycling microscopic community, resulting from climate and ecological changes, where observations on positive methane flux increases were made. Methane-cycling microbes play a crucial role in determining the amount of this gas, which is recycled back into the ecosystem.

Introduction to Methane-Cycling Microorganisms: Soil-exchange of methane together with the environment is determined by two forms of microorganisms, which are methanotrophs and methanogens. The divergent ecological requirements of these different groups, primarily oxygen levels, temperature, moisture content and nutrient accessibility, determine the overall methane flux of any particular ecosystem. Methanogenic archaea, which is particularly active in anaerobic settings, produce methane as a metabolic waste which makes the primary biological source of methane in landfills, agriculture and natural systems. Methanotrophic bacteria thatare sometimes called methane-oxidizing bacteria are active mostly in aerobic environments and receive energy and carbon compounds through methane oxidation.

- Methanogens: In natural ecosystems, methanogens generate around 33 % of emissions. Most anthropogenic methane emissions coming from waste-management and farming, also play a role in the reaction of methanogens. A high number of methanogens are classified as anaerobic archaea, plus there still exists a diversity of methanogens which loosely fit into the 2 main non-phylogenetic groups, which are hydrogenotrophic varieties that produce methane mainly using hydrogen isotopes and carbon dioxide; Including those which are acetotrophic, that is, they primarily use acetate for nutrient breakdown that has occurred from previous matter disintegration activities. The majority, if not all, recognized methanogens contain an MRT isozyme, commonly referred to as methyl-coenzyme M reductase. The genome which encodes α subunit compound is found in nearly all known methanogens.
- Methanotrophs: It's the most widespread class of methane consumers, which works aerobically. Methanotrophs are commonly present in oxic soils, or microsites present around anoxic soils. They are the only recognized biotic sink for methane, being key microbes within the soil community that directly derive energy from methane conversion to CO_2. Methanotrophs form part of the sub-group of methylotrophs that also features methanol oxidizing microbes. Generally, there are 12 genres of methanotrophs which are phylogenetically subdivided into type I consisting of Gamma proteobacteria, and type II within the group Alpha proteobacteria. Some key enzymes need for methanotrophic activities to take place are; monooxygenase (MMO) and methane. With the former occurring as both soluble (sMMO) and particulate (pMMO) forms.

Microbial effects on methane flux: The ability to generate or consume methane is spread among comparatively few microbial taxa, which are phylogenetically unique. The slim distributions of these factors imply that methane production and intake rates are closely tied to the microscopic community structure and abundance compared to any other biogeochemical processes.The genes responsible for methane-cycling are based in deep-branching microscopic clades, same to other complex microscopic traits like sulfate reduction and oxygenic photosynthesis. In contrast, genes responsible for heterotrophic assimilation of other carbon elements may not be extremely conserved (West et al, 2015).

9.8.3 Microbial Communities and Nitrogen Cycle

Nitrogen constitutes around 70 % of the earth's atmosphere, which makes it the most abundant gas in the atmosphere. However, due to escalating anthropogenic nitrogen contribution, numerous aquatic systems experience variations in the nitrogen-cycle which could cause eutrophication in some ecosystems. Studies show that around 90 % of most nitrogen input is minimized through different processes, like nitrogen fixation, denitrification and anammox. The changes between different nitrogen traits are driven by microscopic processes, which occur simultaneously in varying sections of the environment, plus are controlled by various ecological factors. Key among them is microbes, which play a vital role in nitrogen transformation procedures (West et al, 2015).

The nitrogen cycle: It refers to the transfer of nitrogen gas between the soil and the atmosphere. The process covers a series of actions that transform nitrogen into organic compounds, which are then recycled back into nature. This is a continuous cycle preserved by decomposers and other microbes. The nitrogen cycle may be subdivided into different reactions, and microorganisms play a crucial part in all these actions.

Biological nitrogen fixation: The breakdown of nitrogen gas to form ammonia is quite energy-demanding. It needs around 16 ATP, including a complex array of enzymes for breaking the bonds such that the nitrogen may mix with hydrogen. Fairly few bacteria though are capable of carrying out this process. Fixed nitrogen is availed to plants from the death and recycling of free-living nitrogen-fixation microbes, or the symbiotic relation of some nitrogen-fixing microbes with plants (West et al, 2015).

Forms of nitrogen-fixing microorganisms: Some nitrogen-fixing microbes are free-living inside the soil, where they fix nitrogen independently from other organisms. Other nitrogen-fixing microbes, on the other hand, develop symbiotic interactions with plants. Such as root-nodulated legumes, like beans and peas, which have rhizobium.

Oftentimes, free-living rhizobia enter the legume via infection lines developed in the plant's root hair. This infection thread is developed by the plant's root cells and not bacteria, plus is formed just in reaction to contagion. The infection thread develops from the root hair cells before penetrating other adjacent root cells, usually with some thread branching involved. Thereafter the root cells proliferate to develop a root nodule (West et al, 2015).

After just a week of contagion, tiny nodules would visible to onlookers with each root-nodule packed with 1000s of living Rhizobium microorganisms called bacteroids. Similarly, root-nodulated non-legumes, consisting of a varied range of woody species like alder are prone to these factors.

Adapting to their ecosystem: Nitrogen-fixing microorganisms contain a catalyst complex known as nitrogenase, which fast-tracks the transformation of nitrogen into ammonia. The substance supplies hydrogen ions, including energy from ATP. Nitrogenase complex is quite sensitive to oxygen and may become inactivated in case of any exposure to it. However, this is not a concern for free-living anaerobic microbes Clostridium. They have a range of mechanisms for shielding the nitrogenase complex, such as increased metabolic rates and physical barriers. Pathogens such as Azotobacter, for instance, overcome this issue by having a greater rate of respiration comparative to other organisms, thereby maintaining lowered ratios of oxygen within their cells. Essentially, microbes play a crucial part in the synthesis of nitrogen to make it useable by plants.

Chapter 10 Breeding For Soil-Borne Pathogenic Disease Resistance

CONTENTS

10.1 INTRODUCTION

Plant resistance is the relative amount of heritable qualities that influence the ultimate degree of damage done by aninsect. Host plant defenses, plant nutrition and community composition (most plants exist in communities) affect the rates at which plants are attacked by pests and diseases. There are different hypotheses on plant resistance. The repellent plant hypothesis shows that less palatable neighbors can fight off herbivores from plant patches. A plant species can discharge evaporative chemical stimuli that affect host-finding ability of phytophagous insects searching for other species. The attractant decoy hypothesis shows that plants of higher acceptability may draw pests away from other plant species. Pests can be lured away from crops of economic importance by trap crops. Associational resistance is mediated by the natural enemies of the herbivores. Pest density on susceptible plant species can be reduced by associating herbivores-susceptible plant species with herbivore-resistant plant species. Chemical and physical interference of neighboring plant species, greater abundances and diversity of natural enemies acts as a contributing factor and host density is predicted to have strong effects on specialist-resource concentration hypothesis.

10.2 ECONOMIC SIGNIFICANCE OF PATHOGENIC DISEASES

10.2.1 Killing Plants

Plant pathogens cause necrosis which is degeneration of the protoplast followed by death of the tissue or organ or plant. The necrotic areas are generally differentiated by the presence of dark brown border around them (Hussain, 2015). Plesionecrosisisthe necrotic symptoms expressed before death of the protoplast . Holonecrosis are the symptoms expressed after the death of the protoplast. The affected tissue turns brown in color e.g. rots and spots. Below are some of the forms of necrosis that cause a negative economic significance.

- Die back is the extensive necrosis of shoots from top/tip to downward e.g. die back of chili by Colletotrichum capsici.
- A blotch is a dark thread of fungal pathogens that can be seen on the surface of spot or a necrotic area covered with brown fungus mycelium on leaves, shoots and stems e.g.The purple blotch of garlic.

- Cankers occur as dead areas in the bark or cortex formation of sharply delineated, dry, necrotic, localized lesions on the stem due to death of cambium tissue of bark e.g. The citrus canker by Xanthomonas axonopodispv. citri. Infection occurs primarily through the stomata, other natural openings and wounds.
- Blight gives a burnt appearance and is a disease characterized by rapid and extensive death of plant foliage. For example, late blight of potato by Phytophthora infestans.
- Blast is the sudden death of unopened bud or inflorescence e.g. Rice blast disease caused by Magnaporthegresia.
- Rot diseases are characterized by plant decomposition and putrefaction. Fungal or bacterial infection can lead to the subsequent discoloration, softening and even decay of succulent plant tissue.
- Downey mildew is a formation of superficial cottony hyphal growth with spores generally on the underside of the leaf. It is caused by a group of oomycetes, e.g. Downey mildew of grapes by Plasmophoraviticola.

These are diseases that cause plant death and thus affect the investment return to the farmers hence, posing a huge purpose in the economic significance of plant pathogens (Hussain, 2015).

10.2.2 General Stunting of Plants

Stunting is the dwarfing and loss of vigor. Using sugarcane as an example, Ca. Phytoplasma causes Grassy shoot disease. It causes severe damage to the ratoon crop. The symptoms include; profuse tillering and grassy appearance of shoots. Shoots growing from diseased setts remain stunted. The leaves are narrow and small like grass leaves. The canes are thin with short internodes, giving a bunchy or grassy appearance to the clump. The leaves of the infected plants end up appearing white or creamy yellow due to their inability to produce chlorophyll. The leaves appear yellowish and, in some cases, may be entirely devoid of any pigment. The lateral buds sprout before the normal time on growing cane (Hussain, 2015).

It causes economic loss due to; its reduced germination of setts from diseased canes and stunted growth and poor juice quality. The threshold level for yield reduction is 15% infection.

Figure 10.1: Stunted leaves of a tobacco plant around other healthy leaves.

Source: http://ephytia.inra.fr/en/C/10845/Tobacco-Stunted-plant-growth

Ratoon stunting disease is of sugarcane is another disease that is also one of the major causes of degeneration or deterioration of certain commercial varieties of sugarcane. Economic loss is caused by poor germination, extra cultivation cost in the removal of weeds, low yields of canes and loss of ratooning capacity of the cane. The extent of the loss depends on the cultivar and its growing conditions. Planting of cuttings that are infected and using contaminated cutting equipment are two of the most common ways to spread disease. Bacterium can be spread over a long distance even after cutting, due to how highly contagious the bacterium is. The system of recirculating planting machines can also carry the bacteria and spread the disease (Hussain, 2015).

10.2.3 Damage of Leaf Tissues

Soil-borne pathogens inflict injury to plants in their attempts to secure food. This could be directly or indirectly. Feeding habits of insects fall into two main groups; chewing insects and sucking insects. Insects that chew off external plants part, grind them up and shallow them and they possess biting type of mouthparts and cause damage. Feeding on the growing points of plants causes retarded growth. They feed on the leaves causing reduction in the leaf surface area and thus hinder growth by affecting the leaf's function in growth such as photosynthesis e.g. Semiloopercaterpillar. The ash weevil notches the edges of the leaves. Flea beetles make small holes in the leaves by feeding. Injury can also be caused by sucking and piercing insects. These insects remain outside and with their mouthparts pierce through the epidermis and suck the sap. Most sucking insects attack the leaves. Aphids cause general chlorosis and ultimate withering and drying of the affected portion. Feeding can also cause yellow speckling of leaves. Removal of cell contents below the epidermis may cause the leaf surface to become white or silver e.g. typical damage caused by thrips on paddy. The main indicators of damage are; discoloration, drooping, wilting, leaf spots and honeydew.

Injury can also be caused by internal feeders. They remain in the plant tissue during most or all of their destructive stages causing damage to the plant within. The internal feeder can be grouped as borers, worms or weevils, leaf miners and gall insects (Katan, 2017).

Leaf miner larvae live in between the two epidermal layers of the leaves and feed on internal content. Some examples are the citrus leaf miner, the cashew and mango leaf miner and the buprestid leaf miner, etc. Affected leaves do not perform their functions such as photosynthesis, transpiration and even food storage in some crops like lettuce and cabbage. This causes economic loss to the farmers depending on the crop that is being grown.

Insects can also cause injury to leaves by laying eggs e.g. Grapevine stem girdler, use of plant parts for making nests e.g.the leafcutter bee and lastly, insects as disseminators of plant diseases e.g. aphids and thrips.

10.2.4 Damage to the Reproductive Organs including Fruits and Seeds

Pests damage fruits and seeds by chewing on then as a source of food or sucking sap from them. Below are some. The larvae of Helicoverpaarmigera damage the ear heads of soughum by partially eating the grains and giving a chaffy appearance. Fruit sucking moths cause the premature fall of fruits as in citrus. Worms and weevils bore into the buds of flowers and their fruits. The larvae may also bore into the bolls, nuts, fruits or the seeds inside the capsule (cotton bollworm, Helicoverpaarmigera). Larvae of insect pests of legumes put their heads into the pod and eat away the seeds inside the pod. Scale insects suck on sap drawn directly from the plant's vascular system causing yellow speckling on leaves. One of the negative effects on fruits is; premature shedding of developing fruits – adults and nymphs suck the sap from twigs, branches and fruits, resulting in dropping of premature fruits and drying of branches (Katan, 2017).

10.3 IMPORTANCE OF PLANT RESISTANCE TO SOIL PATHOGENS

Harmful soil pathogens such as Salmonella, Escherichia coli, and Listeria are commonly found in farms and gardens where they can cause devastation to both flora and fauna. However, when a crop develops resistance to these microbes, they won't be able to spread destruction within the surrounding area. Below are some of the main benefits of plant resistance to microbes:

10.3.1 It prevents the Spread of Soil-borne Plant Diseases

When a plant is affected by a pathogen, there's little that can be carried out to reinstate its health. In fact, such crops end up being carriers of infection that can potentially be transferred to other parts of the field. For viruses, in particular, they are usually transmitted within their host cells systemically once carried from the soil. Some can be transferred by grafting buds or branches from diseased crops onto healthy ones. Similarly, other disease-causing viruses may be carried or transmitted naturally through insects and mites found in the soil, which are commonly known as vectors (HussainN, 2015).

The main virus-carrying insects consist of around 200 types of aphids living in soils, which transfer mostly mosaic viruses, including more than 100 kinds of leafhoppers carrying the yellows-type viruses. Some of these pathogens may remain infectious for several weeks or months or even replicate themselves inside their insect vectors; while others can only be carried for less than 1 hour after vector contact. No matter the method of transmission, if a plant develops immunity against such pathogens, they won't be able to use it as a host.

A number of soil-borne plant viruses are known to science, they are responsible for causing tobacco rattle, grape fanleaf, tomato ring spots and various strawberry viruses. Mostly, these are spread through nematodes that feed externally on crop roots. Though, some might be transferred by the spores of primeval, soil-inhabiting pathogenic fungi, like those responsible for causing tobacco necrosis and soil-borne wheat mosaic. Plant pathogen resistance also works for viruses which usually overwinter within biennial and perennial plants, including crop debris and insect vectors. It prevents perpetual infection since plants once infected, will normally remain that way for life (Hussain, 2015).

10.3.2 Helps to Control Soil pH

When pathogens multiply in huge numbers, feeding off their host plants, they may eventually interfere with the soil conditions by making them too alkaline or acidic for certain plants to grow in. Therefore, soil microbe resistance is needed to defuse excess toxin concentrations, and prevent pH levels of the nurturing medium from mounting to unacceptable standards.

Spraying plants with pesticides are also not recommended since apart from removing pathogens, fungal spores and insect pests, they also affect the soil pH with their chemical ingredients. Applying a temporary spray during the evening is even more dangerous since it results in slower drying time, due to cooler temperatures as well as lack of sunlight. This allows more chemicals to enter the soil and alter its pH condition (Hussain, 2015). Moreover, even though applying spray pesticides may benefit the crop's health temporarily, foliage which remains wet for extended periods of time is prone to infections that require humid conditions to grow. To solve this problem, crop-resistance methods against pathogens can be practiced, where no toxic chemicals are used to control the microbes that may otherwise affect the plant also in a negative way.

10.3.3 Prevents the Spread of Diseases to Humans

Some harmful bacteria and fungi present in the soil have been known to carry infection to humans. For instance, there have been fatality reports coming from Legionnaires' disease, which is a lung infection attributed to soil-borne bacteria. Many pathogens that cause illness in people reside in the soil, so it's not shocking that plants that don't have resistance to these disease-causing microbes may act as hosts for their multiplication.

Soil-borne pathogenic infections may lead to sepsis, whereby the microorganism enters the person's blood and steadily grows in population, causing the patient's body to react with an inflammatory effect that may lead to organ failure or septic shock if treatment is not found quickly. Individuals with low immunity levels, like those with infections such as lymphoma or HIV, are generally at a higher risk of catching illness from soil-based pathogens than others. Even typical daily activities, like weeding the garden or pruning, can expose them to harmful pathogens lurking in soil residues. So, it's even more imperative that they use microbe-resistant plants in order to reduce their chances of being affected by opportunistic soil-borne illnesses (Hussain, 2015).

Signs and symptoms of infection from contaminated soil usually depends on where the disease is, including what particular microbe is causing it. Otherwise, some of the symptoms patients of microbe contagion may initially experience are fever, aches and soreness, which may either be moderate or severe. Also, if the infection affects the person's respiratory tract, then coughing, shortness of breath, including difficulties with breathing may also occur.

10.3.4 How Pathogen Plant-resistance can benefit the Farmer

Pathogenic microorganisms found in agricultural soils may have a disastrous effect on crops by inducing; resistance to crop control products, pathogenicity, poor soil nutrition and reduced fertility, low yields or even crop loss. These microbes are usually difficult to manage or control, once widespread. Meaning there's very little a farm owner can do if they reproduce profusely.

The easiest way to prevent such a scenario is by using microbe-resistant plants in your field, these will constrain the pathogens by denying them appropriate conditions for reproducing and spreading all across the field. Some pathogenic microorganisms, if not controlled, will decay root nodules, discharging nutrients from the crops into the soil and providing conditions for weeds to grow. Additionally, some microorganisms reduce the efficacy of nutrient uptake or mobilization, further triggering nutrient deficiency and slowed plant growth (Hussain, 2015).

Likewise, the presence of pathogens may impact photosynthetic functions, leading to leaf or stem chlorosis and necrosis, which causes a reduction in the transmission of moisture and nutrients across the plant's vascular system.This reaction often occurs with the aid of pathogen-released toxins thatcolonize the host plant's tissue, consequently leading to stunted growth or even other conditions like blight, canker, leaf scorch or gummosis.

10.4 HISTORICAL BACKGROUND FOR BREEDING FOR PATHOGENIC DISEASE RESISTANCE

There are three methods of plant genome manipulation; classical breeding, targeted genome editing and transgenic approach. This process involves the altering of DNA in an organism's genome. In 1991, Dewit et al reported the first fungal avirulence gene avr-9 from Cladosporium fulvum. In 1992, Johal and Briggs reported the first R gene HM-1 in maize, resistant to Cochlioboluscarbonum race-1. In 1994, FlavrSavr reported the first transgenic tomato in USA.

The method for gene transfer was established;

- Identification and cloning of the gene of interest. Public and private labs are directing more efforts into locating, identifying, characterization and cloning genes of agricultural importance.
- Adding DNA segment to initiate gene expression. Promoter: CaMV 35S promoter. Terminator sequence: Nopaline synthase transcription terminator sequence from Agrobacterium tumefaciens.
- Adding a selectable marker. A gene transfer marker is needed to identify cells with the new genes. Typically, antibiotic resistance genes are used as markers.
- Gene transfer.

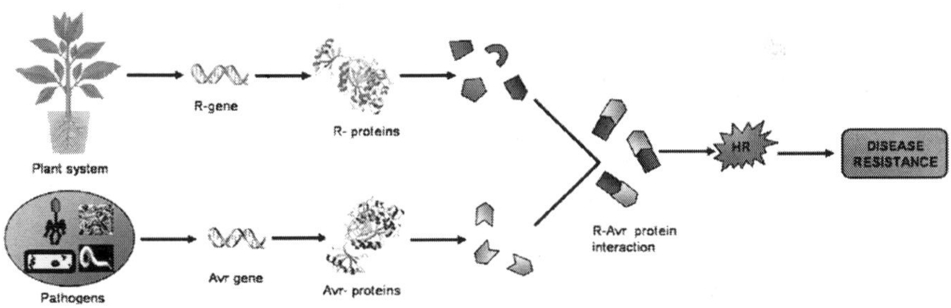

Figure 10.2: Plant-pathogen interaction and development of disease resistance.

Source: https://www.researchgate.net/figure/Plant-pathogen-interaction-and-development-of-disease-resistance_fig1_258021859

Properties of a good host were noted to be; easy to transfer, support the replication of the recombinant DNA, free from elements that interfere in the replication of recombinant DNA, lack active restriction enzymes, and do not have methylases.The mechanisms of transgenic plants in disease resistance are; directly toxic to or reduce growth of the pathogen, destroy or neutralize the pathogen components, potentially enhance the structural defenses in the plant and release resistance gene products involved in R/Avr interactions and hypersensitive response.

The steps in breeding for disease resistance are; sources for resistance —> methods of breeding for disease resistance —> testing for disease resistance —> selection for resistance lines.

The sources for disease resistance are; known variety, wild relatives, related species, mutations and unrelated organisms.

Advantages of breeding for disease resistance are; resistant varieties offer the cheapest means of disease control, resistant varieties obviate the use of fungicides, thus reduce environmental pollution, effectiveness of resistant varieties is not affected by environmental conditions, it safeguards against the inadvertent release of such varieties that are most susceptible than earlier varieties and in rusts and smuts it is the only feasible approach for disease control (Hussain, 2015).

10.5 CATEGORIZATION OF RESISTANCE TO PATHOGENIC DISEASES

10.5.1 Susceptible Reaction

This is found in biotrophic pathogens (obligate parasites) and is associated with the hypersensitive response of the host. Only one of the four combinations would lead to the resistance response since the products of R and Avr would recognize and interact with each other. The product of alleles a and r are unable to recognize each other and there is no interaction between them, hence the reaction of the host becomes susceptible, which is an incompatible reaction.

According to Francisco et al (2001), susceptibility is the main factor for specificity in the gene for gene relations. He explains it with the help of interactions of five hosts and five pathogens attacking them specifically. Suppose there are five host varieties with five different R genes; R1, R2, R3, R4 and R5. A pathogen having virulence gene V1 attacks a plant with the resistance gene R1 and not a pathogen without this resistance gene regardless of how many the virulence genes it may have.

Table. The diagonal check for specificity in a gene-for gene relationship [a]

Pathogen	Plant				
	R_1R_1 [b]	R_2R_2	R_3R_3	R_4R_4	R_5R_5
v_1v_1	S	R	R	R	R
v_2v_2	R	S	R	R	R
v_3v_3	R	R	S	R	R
v_4v_4	R	R	R	S	R
v_5v_5	R	R	R	R	S

a. Plant reaction when resistance gene R_1, R_2, R_3, R_4, R_5 at five loci interact with virulence genes v_1, v_2, v_3, v_4, v_5 at five loci in the pathogen

b. Resistance is assumed to be dominant and RR can be replaced by Rr. Virulence is assumed to be recessive. However, recessive resistance and dominant virulence are also known.

R= resistant S= susceptible

Figure 10.3: Gene for gene relationship.

Source: https://www.slideshare.net/KAUSHIKKUMARPANIGRAH/presentation-on-gene-for-gene-hypothesis-and-concept-of-vertical-and-horizontal-resistance

It may be expected that when a new variety of carrying VR is developed, it is resistant to the prevalent pathotypes of the pathogen in question. It, therefore, becomes popular rapidly and occupies a large proportion of the growing area under the concerned crop. This period of boom continues until the pathotypes specific to the VR gene employed in the variety evolves and/or increases in frequency to cause an epidemic in the field occupied by the variety. The now susceptible variety loses popularity and the area under it declines sharply thus the variety is effectively busted by the virulent pathotypes. This is the boom-bust cycle by Priestey (1970). When a resistant variety becomes popular, it is cultivated in large areas. The plants of this variety act as strong selection agents for the pathotypes specific to the new variety since only this pathotype can cause disease in the plants of this variety.

10.5.2 Resistance Reaction

This occurs in heterotrophic pathogens (facultative parasites). The allele for the susceptibility of the host (r) and those for virulence in the pathogen produce specific compounds, which interact with each other to produce susceptible response. One of the four combinations would lead to susceptibility and the rest lead to resistance.

Figure 10.4: Image showing hypersensitive reaction seen in Vertical resistance.

Source: http://www.helsinki.fi/ppvir/research/nova/Hofius.pdf

The hypersensitivity reaction is responsible for limiting the growth of the pathogen, inthis way, it provides resistance to the host against the pathogen. Most involve only a single cell or very few cells. Whenever the virulent pathogen is artificially injected to a non-host plant or resistant plant and when a virulent strain is injected into a susceptible host plant, it induces a hypersensitive response. The HR is the point of the highest development of the plant defense responses and is started by the recognition of specific pathogen signal molecules, known as elicitors by plants. The hypersensitivity response occurs only in specific host-pathogen combination in which the host and pathogen are incompatible. This occurs only in vertical resistance (Hussain, 2015).

Van der Plank introduced the Vertifolia effect which was from the German potato variety 'Vertifolia' having the late blight resistance genesR3 and R4. The variety became susceptible when the appropriate pathotypes P(3,4) evolved in the pathogen leading to a nearly complete failure of the crop; such a total failure of vertical resistance leading to a disease epidemic called Vertifolia effect.

This reaction can be used to; develop resistant variety against obligate pathogens, as a source of pathogenic variability in pathogens, mutability of resistance and virulence genes, prediction for putative genotypes, to see why host resistance is expressed under one set of conditions and not others and for cataloging and storing of R genes in the form of plant seeds or cutting and V genes in the form of pathogenic strains.

10.5.3 Tolerance Reaction

Tolerance to diseases is the capability of plants to create a good crop despite the fact that they are infected with a pathogen. It results from specific characteristics of the host plant that permit the pathogen to evolve and multiply in the host while the host and still manages to produce a good crop. They are susceptible to the pathogen but are not killed by it and show little to no damage. Some of the plant tolerance mechanisms are; altered gene expression, adjustment of hormone balance, synthesis of stress proteins, activation of signaling factors, enhanced antioxidative metabolism, facilitated membrane transport, etc.

Plant reactions to an invading pathogen center onthe hypersensitivity reactions. The hypersensitive reactions initiate changes in plant physiology and biochemistry: early activation of defense-related genes to synthesize pathogenesis- related proteins; protease inhibitors to stop cell wall lysis by specific enzymes expressed by pathogen and bacterial cell wall lytic enzymes (chitinase and glucanase). Changes in cell wall composition; expresses enzymes providing structural support to cell walls via synthesis of lignin, suberin, callose and glycoproteins. It causes apoptosis at the invasion site of the plant to physically cut off the rest of the plant. Synthesis of secondary metabolites to isolate and limit the pathogen spread (isoflavonoids, phytoalexins) (Hussain, 2015).

Pathogens carry avirulent genes and hosts carry resistance genes. The presence of both prevents pathogens from attacking the plant. Infection occurs when pathogens lack avr genes or plant is homozygous recessive for resistance genes.

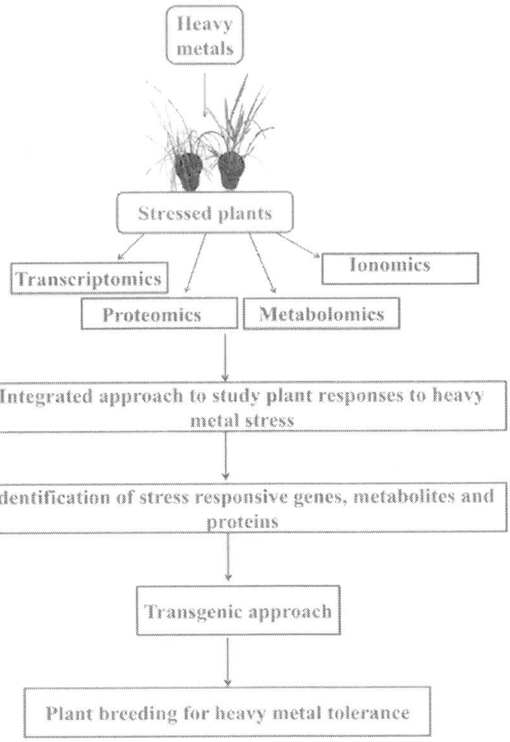

Figure 10.5: Image shows tolerance to heavy metal. This process is also followed in tolerance to pests.

Source: https://www.frontiersin.org/articles/10.3389/fpls.2015.01143/full

The long-term plant resistance to a pathogen is the systemic acquired resistance. Secondary metabolites induced by the hypersensitive reaction-initiated change in metabolism in other plant organs through the control of signal transduction chains. Salicylic acid both induce pathogenesis- related gene expression and enhances resistance to infection by plant viruses (Katan, 2017).

Local salicylic acid production induces distal production and systemic acquired resistance via; Salicylic acid transport in the xylem and methylation into MSA, volatilization and distal detection. High constitutive salicylic acid levels resulted in plants with a high ability to resistpathogens.

10.6 VERTICAL AND HORIZONTAL RESISTANCE

10.6.1 Vertical Resistance

Disease-resistant variations of plants provide an effective, safe, and moderately affordable way of controlling many crop syndromes. Many available commercial strains of crop plants have resistance to one or more pathogens. Normally, in vertical resistance, crop varieties display a high amount of resistance towards a single strain or race of pathogen. This unique ability usually determined by one or afew plant genes. Resistant or immune gene varieties are critically significant for low-value plants whereby other controls are unattainable, or their cost makes them unfeasible. Much has been done in developing illness resistant strains of field crops, fruits, vegetables, ornamentals and turf grasses. But more flexibility and capacity for genetic change persists in most economically viable plants, plus pathogens are becoming flexible as well. Sometimes, a new crop variety is established that's highly vulnerable to a previously insignificant pathogen (Francisco et al, 2001).

Figure 10.6: A plant breeder inspects a young crop to determine if it's suitable for disease-resistance crossing.

Source: https://www.toppr.com/guides/biology/strategies-for-enhancement-in-food-production/plant-breeding/

Characteristics: Vertical resistance is exhibited in a wider range of environmental conditions such as temperature, humidity and light. It features more concentrated inoculum, plus in a lengthier part of the plant life cycle compared to horizontal resistance. Case in point, crop selection for the vertical resistance towards cereal leaf microbes, such as mildew fungi and leaf rust, is typically done in the sprouting phase which doesn't have any horizontal resistance. Vertical resistance DNAs may also differ in reaction to the injection conditions. For illustration, flax has numerous loci where rust preventive genes are mapped. For locus L, resistanceL1, L2 and L3 offer only a pair of alternatives based on the race applied for inoculation, total immunity or susceptibility, and reaction of gene N2 varies based on the illumination. Considering at vertical resistance, among the greatest factors in trait manifestation is achieved through interacting genes, whereas the function of additive or environmental influences is minimal, the phenotypic variation may display characteristics whereby vertical resistance protects from some types of blight fungus and not others (Francisco et al, 2001).

10.6.2 Horizontal Resistance

Horizontal resistance, contrarily, protects crop varieties against multiple strains of a disease-causing microbe, though the protection is not as holistic as vertical. Moreover, horizontal resistance is widely common and comprises many genes. For genetics, the phrase horizontal resistance was originally used by researcher J.E. Vanderplank, who coined it to describe multiple-gene resistance, which may sometimes be called generalized resistance as well (Francisco et al, 2001).

Another scientist, Raoul A. Robinson, further explained the meaning of horizontal resistance. Unlike vertical resistance or parasitic capacity, horizontal resistance and parasitic ability are fully independent of one another in the genetic definition. During the first phase of horizontal resistance breeding, crops are exposed to microbes and chosen for partial resistance. But those with minimal resistance perish, and crops unaffected by the microbe have vertical resistance meaning they are removed. Consequently, the remaining flora haspartial protection and their seed is kept and bred back again to sufficient quantities for further testing.

With the goal that in the remaining crops there could be several strains of partial-resistance genes available, and by crossing this pool right back on itself, several half resistance genes will assemble and provide improved resistance to a wider variety of pathogens.

Successive breeding-rounds for horizontal resistance occur in a rather traditional fashion, where choosing plants for illness resistance is measured by produce. The crops are exposed to local regional pathogens, plus given minimal support towards fighting them. This form of resistance is polygenic, meaning it acts against multiple races of blight. In fact, vertical resistance, provided by the R-genes in plants is oligogenic, though it can be overcome through a change of race(Francisco et al, 2001).

Generally, horizontal resistance slows the momentum at which infection spreads across the plants in field. It also helps to activate the R-genes, which might have gone into disrepute after being usedwrongly. That is if they're applied in replacing horizontal resistance, while should ideally be used just to supplement it.

10.7 SYSTEMIC ACQUIRED RESISTANCE

Systemic acquired resistance, or simply SAR, is a holistic plant resistance reaction that happens following a previous localized contact to microbes. SAR is comparable to the inherent immune system present in animals, plus there's evidence to show that SAR in crops and natural immunity in creatures may be evolutionarily preserved.

Additionally, SAR is vital for crops to resist infection, as well as for recovery from sickness once formed. It can be induced through a wide variety of pathogens, particularly those which cause tissue necrosis, plus the resistance seen following the introduction of SAR is effectual against a variety of pathogens, this is why SAR-resistance is commonly referred to as "broad spectrum". This breeding method has been witnessed in a wide array of flowering plants, such as monocotyledon and dicotyledonous species. It can be used in corn, though the commonly used commercial substance benzothiadiazole might not be effectual against Puccinia sorghi that results to common rust (Hussain, 2015).

Crops use pattern-identification receptors to distinguish conserved microbial signatures. The recognition prompts an immune reaction. Plants also transmit immune receptors thatrecognize highly variable microbe effectors, which comprise the NBS-LRR group of proteins. SAR has been associated with induction towards a wide variety of genes, which are the pathogenesis-related DNAs, plus the stimulation of SAR actually requires the buildup of endogenous salicylic enzyme. The microbe-induced SA signal stimulates a cellular signal transduction passage that's recognized by a gene known as SAI1, NIM1 or NPR1 which are 3 names describing the same gene. In essence, systemic-acquired resistance is a type of induced resistance that's activated all around a plant after it's gotten exposed to elicitors coming from avirulent, virulent and nonpathogenic microorganisms, or artificial molecular stimuli like salicylic acid (SA) and chitosan. Stimulation of SAR is marked by the increment of salicylic acid for stimulating defense mechanisms, which often results in a localized hypersensitive reaction.

10.8 GENETICS OF DISEASE RESISTANCE

According to plant evolution studies, resistance to some infectious diseases can be transmitted genetically from a parent crop to its offspring. This genetic resistance is beneficial for species survival in the midst of devastating disease outbreaks since those having this gene are less likely to encounter the infections or spread it to other plants around them. Scientists used empirical models to estimate the possible consequence of 'genomic selection' on natural evolution. While natural selection is classically regarded as 'survival for the fittest', in the case of plant genetics it may be more precise to assert that its 'survival for the fittest families.' Modern studies on this topic have resulted in the following conclusions:

- If a plant relative has a genetic predisposition that provides resistance against illness, it shall benefit other relatives since they would not likely contract the disease.
- Genes that give resistance to contagious diseases usually cluster within particular families.
- Hence, the resistance genes found across a family assist each other out, plus natural selection acting in their favor might be drastically increased.

This model could prove useful in comprehending the spread of lethal diseases, and further, alter the future natural selection of some genes within a given population. Researching the genetic traits of these infections could be a crucial step towards, getting more insight intothe evolutionary cycle of infectious illness resistance. Moreover, recovery from infection depends on the effectiveness and protective abilities of the immune system. Identifying genetic material that affects immunity may assist in identifying plants with quicker or more advanced responses to illness challenges. Crop geneticists began addressing genome disease resistance by evaluating whether immunity traits were largely heritable, afterward they mapped the DNAs involved which are known as quantitative trait loci or just QTL. The research confirmed that the percentage of several immune cell subsets and plant serum antibody amounts were heritable in plants (Hussain, 2015).

10.9 THREE MODES OF INHERITANCE

10.9.1 Oligogenic Inheritance

Oligogenic inheritance is defined as a trait thatis influenced by multiple genes and not just one. It represents a transition between monogenic heredity, whereby a trait is influenced by just one causative genome and polygenic heredity in which features are determined by multiple genes and usually environmental factors as well. Originally, many traits were believed to be determined by one causative gene in what arecalled monogenic inheritance, but work in genetics showed that these characteristics are somewhat rare, plus in most cases the so-called monogenic aspects are chiefly influenced by a single gene, though can be mediated using other genes of lesser effect.

Modifier genes: An example of oligogenic heredity is a case whereby one gene is enough to trigger a trait, but its infiltration or expression is determined by another gene, known as the modifier. Modifier genes may not be useful by themselves, but in the presence of another stronger DNA, they greatly improve their function.

Identifying oligogenic traits: A trait may be identified as oligogenic after checking the below lines of evidence:

- Phenotype–genotype connections: If a phenotype can't be projected to a single solidly correlated locus, rather the addition of genome-class from another locus enhances the correlation, which is clear evidence that the trait is oligogenic.
- Phenotypic differences in a plant model of the characteristics which are reliant on the genetic history. Here, the outcomes of a probable modifier locus may be tested in a plant with another recognized causative mutation.
- Variances between mutations and the Mendelian system of inheritance, which follows the policy of segregation and autonomous assortment, whereby a gene transmitted from either parent crop subdivides into gametes at a similar frequency. If mutation carriers don't display the form of phenotypes projected under Mendelian inheritance, then other models could better explain detected patterns of inheritance. Also considered for oligogenic traits, is the establishment of linkage to several locior failure to distinguish association from Mendelian models.

10.9.2 Polygenic Inheritance

Polygenic inheritance points to one characteristic that's determined by multiple genes, also known as multifactorial inheritance. Its patterns usually follow a unique bell-shaped distribution format, showing continuous variation. Through increasing the number of genes that control a trait, the overall number of phenotype groupings also increase, up to the number of phenotypes by which a plant can be allotted are no longer separate, but continuous (Hussain, 2015).

Qualities of polygenic traits: The phrase polygene was first coined by Mather, a genetic scientist, in 1941. Since then, it has widely been applied in quantitative genetics substituting the older phrase 'multiple genes.' Some of the key attributes of genetic polygenic inheritance are as below:

- Each polygenic feature is determined by various autonomous genes and each one has cumulative effects.
- Polygenic attributes exhibit constant variation instead of a discontinuous variation. Therefore, they can't be categorized into any clear-cut groups.
- The effect of single genes is not easily noticeable for the case of polygenic features and, thus, such traits may also be referred to as minor gene aspects.
- Statistical analysis of polygenic variances is based upon averages, variances and co-variances, while the discontinuous variation is typically assessed with the aid of ratios and frequencies, Therefore, polygenic characters are calculated in quantitative genetics, as well as oligogenic aspects in Mendelian genetics.
- Polygenic features are quite sensitive to ecological changes, while oligogenic characters are minimally influenced by the same environmental variations.
- The categorization of polygenic attributes into distinct clear-cut sets is not feasible due to continuous variation, ranging from a particular extreme to another. As for qualitative characters, these groupings may be possible due to discrete or irregular variation.
- In general, the appearance of polygenic features is determined by additive gene function; however, cases have been reported where polygenic characters have been determined by dominance, including epistatic gene action. As for oligogenic characters, the genetic action is predominantly of non-additive form.

10.9.3 Cytoplasmic Inheritance

Cytoplasmic inheritance is also known as extra-nuclear inheritance refers to the transfer of genes that are found outside the cell nucleus. This hereditary method is commonly found in eukaryotes, including the cytoplasmic organelles found in plant mitochondria and chloroplasts.It tends to differ from ordinary nuclear genetics since it doesn't adhere to the principles of gene inheritance, stating that half of the genes found in an offspring originate from either parent. Rather, based on the particular, DNA present within the organelles may come entirely from the father, mother, or sometimes uneven mixture of both (Hussain, 2015).

Several plant breeders have found evidence that the majority of all the organelle DNA, present in some groups of crops, is directly derived from the mother. For instance, in one research involving 295 different plant species, it was discovered that 238 displayed maternal inheritance qualities. The unclear aspect for plant breeders though is, if all crop cells, including pollen, have organelles, then what exactly blocks the sire's organelle DNA from penetrating the egg? Still more research is needed to understand this particular aspect.

Types of extranuclear inheritances:

- Vegetative separation: This results from unsystematic replication and splitting of cytoplasmic organelles. The process occurs inside mitochondria and chloroplasts for mitotic cell divisions, causing daughter cells with a random match of the original parent cell's organelles. Case in point, vegetative separation of mitochondria occurs asexually in replicating yeast cells.
- Uniparental heredity: It occurs in extra-nuclear DNAs when just one parent provides an organellar genetic inheritance to the progeny. A good example of uniparental genetic transmission is with maternal heredity of human mitochondria. In such cases, the mother's mitochondria aretransferred to the progeny during fertilization through the egg. But the father's mitochondrial DNA is not transmitted over to the offspring through male reproductive cells. Nevertheless, a few rare cases have been described of paternal mitochondrial heredity in some species, which is a genetic oddity.

10.10 METHODS OF BREEDING FOR DISEASE RESISTANCE

10.10.1 Selection

Selective or artificial selection breeding is a process whereby cultivators use plant breeding techniques, to distinctively develop certain phenotypic features by choosing parent plants that exhibit good characteristics for reproduction. When bred in this manner, domesticated plants may be classified as varieties, cultivars, cultigens, or breeds. Two purebred plants of different species produce what's known as a crossbreed, and these plants are characteristically known as hybrids. Vegetables, flowers, or fruit-trees can be cultured by amateur growers, including commercial and non-commercial professionals. But major crops are typically the purview of professionals. Plant species are selected for certain crucial traits thatprevent the spread of diseases.

During breeding, techniques like linebreeding, in-breeding, and outcrossing may be utilized. Selective breeding was first hypothesized by Charles Darwin in his publication titled, Origin of Species which was released in1859. Its initial chapters elaborate selective breeding techniques and domestication of various species. Darwin applied artificial breeding as a way of highlighting how quality attributes, such as disease resistance, can be passed down from one generation to another (Hussain, 2015).

The measured utilization of selective breeding in producing desired outcomes has become quite common in farming and experimental biology. This form of breeding may be unintentional, for example, emerging from the procedure of human cultivation; or it may cause some unintended results. Case in point, in some seeds, a rise in seed size could have emerged from certain plowing activities apart from the intentional assortment of larger seeds.

Most probably, some form of interrelation between natural and manmade factors result in plant selection and domestication. Furthermore, nowadays scientists use selective plant-breeding for research to come up with transgenic plants thatbreed "right" for artificially implanted or deleted genes.

Equally, selective breeding may be an effective way of determining whether a particular trait can develop in reaction to selection. The single-generation technique of breeding is not as precise or direct compared to this method (Hussain, 2015).

10.10.2 Introduction

Plant introduction is the process of trans-positioning crops from their venue of cultivation, over to places where they weren't grown earlier. There are two types of introduction, which are direct and indirect. The introduction of crops is among the earliest techniques used for plant breeding. Mature plants found in one place are transferred to another area for utility. Likewise, quarantine plays a critical role in introduction as it ensures the material that's to be added doesn't carry any seriousinfections with it.

Perspective:This technique is an ancient process of crop improvement. Immediately after America's discovery, various crops such as potato, tomato, peanuts, maize, beans, peanuts and peppers were swiftly introduced to supplement already existing crops. However, to prevent the spread of diseases between different plant species, introduction breeding was applied in farms all across the country. The method is effective since it's compatible with all types of crops, whether cross-pollinated, self-pollinated or asexually reproduced.

Methods of Introduction:In earlier times, the introduction of crops from one venue to another was mostly done by traders, invaders, travelers, and merchants. However, nowadays there are professional plant breeders who are well established in this role. Some of the introduction techniques they use are:

- Primary Introduction:Primary introduction is a technique often used for commercial agriculture, being a variety without any noticeable alteration in the original genome type. There are various crops where primary introduction breeding has been applied successfully, such as wheat varieties Lerma Rojo and Sonora 64. Both of these varieties were derived from Mexico and used directly for largescale cultivation in India. Similarly, in semi-dwarf rice varieties Taichung Native 1, IR 8. IR 20 and IR 36 are examples of primary Introduction. Many such examples can be cited from other field crops and fruit crops.
- Secondary Introduction:In the secondary introduction, a plant is used as breed variety following selection from the first genotype sampled, or used for transferring certain desirable genes to an already cultivated variety.

10.10.3 Mutation

Mutation breeding, also known as ”variation breeding” is a technique of exposing plant seeds to radiation or special chemicals, for purposes of generating mutants with appropriate characteristics to be reared with other cultivars. Crops produced through mutagenesis are also sometimes known as mutagenic plants. Researchers believe that from 1930-2014 around 3200 mutagenic crop species were released into farms, having been gotten either as primary mutants or progenies of the original plant (Hussain, 2015).

The mutation process:There are various types of mutagenic breeding like using chemical mutagens, such asDiethyl sulfate, methanesulfonate or which are applied to produce mutants. This form of breeding is normally used to develop traits in plants like bigger seeds, sweeter fruits or new colors, though nowadays it's also increasingly being used for disease-resistance breeding.

- Radiation breeding: It involves exposing crops to controlled radiation, which is sometimes referred to as radiation breeding. Being a subgroup of mutagenic breeding, it was discovered during the 1920s by Lewis Stadler, a University of Missouri professor who successfully used X-rays on barley and maize plants. For barley, the resulting crops were yellow, white, pale yellow, plus some with white stripes. Soon afterward, radiation-induced mutant strains were developed mainly using gamma rays at concentrations of about 64%, and X-rays at 22%. With advancements in technology, radiation breeding could in the future occur in atomic gardens.
- Chemical mutagens: High degrees of chromosomal deviations ensuing from ionizing radiation, including the accompanyingharmful effects have made scientists look for alternative sources for prompting mutations. Consequently, a variety of chemical gene mutagens have been developed. The most broadly used chemical activators are alkylating agents.

Key among them is ethyl methanesulfonate (EMS), which is highly valued due to its effectiveness and handling ease, including its simplified detoxification that's done via hydrolysis for discarding. EMS has grown to become a widely used mutagen for creating a wide array of mutant species for screening, like in the development of tilling populations.

10.10.4 Hybridization

Hybridization is the crossing of genetically separated plants, and may further involve crossing individual crops within a given species, or pairing individuals between different species. This genetic breeding method is used by agriculturalists, as a way of creating plant forms with disease-resistant traits. Ideally, the ultimate goal of hybridization is producing offspring crops that perform better, compared to their parental species. Hybridization can also be applied to transfer various desirable qualities from one subspecies of plants to another, combining valuable traits from different species into one group, or producing sterile plants (Hussain, 2015).

In the 20th century, calculated hybridization between carefully selected parent plants became central in the reproduction of self-pollinated species. However, since genes are typically in the group of other genes within a genotype collection, a plant breeder's main challenge is to efficiently manage the large numbers of genotypes thatresult in successive generations following hybridization. A good example of hybridization in developing variability, where, for example, a mix between hypothetical wheat strains varying by just 21 genes is able to produce well over 10 billion different genotypes for the crop's second generation. Though the greater majority of the second-generation genotypes are heterozygous for at least one or several traits. It's still statistically probable that more than 2 million different pure-breeding genotypes may occur, with each one potentially having a pure-line variety. Pedigree breeding commences with the pairing of two genotypes, with each having a few desirable characteristics lacking in other genomes (Hussain, 2015).

Additionally, if the original parent pair doesn't have all the required characteristics for disease-resistance offspring, a third parent may be introduced by crossing it together with a hybrid progeny coming from the first offspring. In pedigree breeding, superior genome types are chosen in successive generations, plus a record is kept of the parent–progeny connections. The second-generation progeny usually gets the first opportunity for breeding in pedigree programs.

References

1. Aboulkacem Lemtiri, Gilles Colinet, Taofic Alabi, Daniel Cluzeau, Lara Zirbes, Éric Haubruge, & Frédéric Francis . (2014). Impacts of earthworms on soil components and dynamics. Biotechnol. Agron. Soc. Environ, 18(1), 121-133.
2. Agtmaal, M. v. (2015). Suppression of soil-borne plant pathogens. Thesis submitted in fulfilment of the requirements for the degree of doctor at Wageningen University, PP. 152.
3. Carolina Sarmiento, Paul-Camilo Zalamea, James W. Dalling, Adam S. Davis, Simon M. Stump, Jana M. U'Ren, and A. Elizabeth Arnold. (2017). Soil-borne fungi have host affinity and host-specific effects on seed germination and survival in a lowland tropical forest. Proceedings of the National Academy of Sciences of the United States of America, 114 (43) 11458-11463.
4. Corliss, J. O. (2016). The Ciliated Protozoa: Characterization, Classification and Guide to the Literature. New York: Elsevier, pp.472.
5. Dr. Isha Rana & Dr.S.S.Randhawa. (2015). Impact of Climate variations on plant diseases. State Centre on Climate Change, 1-20.
6. Emerson, J. (2018, July 17). Soil viruses: unlocking the secret garden. Retrieved from naturemicrobiologycommunity: https://naturemicrobiologycommunity.nature.com/users/117403-joanne-emerson/posts/36561-soil-viruses-unlocking-the-secret-garden
7. Emerson, J. B. (2019). Soil Viruses: A New Hope. American Society for Microbiology, 10.1128/mSystems.00120-19.
8. Francisco xavier ribeiro do vale, j. E. Parlevliet, & laércio zambolim. (2001). Concepts in plant disease resistance. 577-589.

9. Gareth Trubl, Ho Bin Jang, Simon Roux, Joanne B. Emerson, Natalie Solonenko, Dean R. Vik, Lindsey Solden, Jared Ellenbogen, Alexander T. Runyon, Benjamin Bolduc, Ben J. Woodcroft, Scott R. Saleska, Gene W. Tyson, Kelly C. Wrighton, Matthew B. Sullivan, Vi. (2018). Soil Viruses Are Underexplored Players in Ecosystem Carbon Processing. American Society for Microbiology, 10.1128/mSystems.00076-18.
10. Graham Stirling, Helen Hayden, Tony Pattison, & Marcelle Stirling. (2016). Soil Health, Soil Biology, Soil-borne Diseases and Sustainable Agriculture: A Guide. Clayton: Csiro Publishing, pp.280.
11. Hoorman, J. J. (2011). The Role of Soil Bacteria. Agriculture and Natural Resources, 1-4.
12. HUSSAIN, B. (2015). Modernization in plant breeding approaches for improving biotic stress resistance in crop plants. Turkish Journal of Agriculture and Forestry, 515-525.
13. Idowu, J. (2014). Principles of Soil Health Management . Extension Plant Sciences Online Training Series, 1-53.
14. Jackie Aislabie & Julie R. Deslippe . (2013). SOIL MICROBES AND THEIR CONTRIBUTION TO SOIL SERVICES. SOIL MICROBIAL DIVERSITY, 143-159.
15. Jenkins, A. (2005). Soil fungi . soil biology basics, 1-2.
16. John Clarkson, Jane Debode, Lorenzo Furlan, Roy Neilson, Ann Charlotte Wallenhammar and Johann Zahrl. (Macromolecules: Indian Journal). Isolation and Identification of Fungi from Contaminated Soil to Build Biological Resource as Biocontrol Activity. SOIL-BORNE DISEASES, 1-4.
17. José M. De la Rosa, Ana Z. Miller & Heike Knicker . (2018). Soil-borne fungi challenge the concept of long-term biochemical recalcitrance of pyrochar. Scientifc Reports.
18. Katan, J. (2017). DISEASES CAUSED BY SOIL-BORNE PATHOGENS: BIOLOGY, MANAGEMENT AND CHALLENGES. Journal of Plant Pathology, 99 (2), 305-315.
19. Lwoff, A. (2014). Biochemistry and Physiology of Protozoa. New York: Elsevier, 1-27.
20. Magdalena Frąc, Silja E. Hannula, Marta Bełka, and Małgorzata Jędryczka. (2018). Fungal Biodiversity and Their Role in Soil Health. Frontier in microbiology, 9: 707.

21. Magdoff, F. (2001). Concept, Components, and Strategies of Soil Health in Agroecosystems. Journal of Nematology, 33(4):169–172.
22. Mark Schonbeck & Michael Stein. (2014). Reducing Risk through Best Soil Health Management Practices in Organic Crop Production. Organic Farming Research Foundation, 1-54.
23. Niall A. Logan & Paul De Vos. (2011). Endospore-forming Soil Bacteria. Berlin: Springer Science & Business Media, 1-89.
24. Patel, D. A. (2016). ADDRESSING SOIL HEALTH MANAGEMENT ISSUES IN INDIA. International Journal of Research - GRANTHAALAYAH, Vol.4 (Iss.12) 1-14.
25. Peter J. Hotez, Donald A. P. Bundy, Kathleen Beegle, Simon Brooker, Lesley Drake, Nilanthi de Silva, Antonio Montresor, Dirk Engels, Matthew Jukes, Lester Chitsulo, Jeffrey Chow, Ramanan Laxminarayan, Catherine M. Michaud, Jeff Bethony, Rodrigo Correa-Oli. (2006). Helminth Infections: Soil-Transmitted Helminth Infections and Schistosomiasis. Disease Control Priorities in Developing Countries, 467-481.
26. Puri, S. (2016). Eco-friendly management strategies for soil-borne plant pathogens. International Journal of Advanced Research in Biological Sciences, 3(1): 69-75.
27. R.N. Doetsch & T.M. Cook. (2012). Introduction to Bacteria and Their Ecobiology. Berlin: Springer Science & Business Media, 14-86.
28. Reddy, P. P. (2011). Biofumigation and Solarization for Management of Soil-Borne Plant Pathogens. London: Scientific Publishers, 1-93.
29. Sukhareva-Buell, N. (2003). Biologically Active Substances of Protozoa. Berlin: Springer Science & Business Media, pp.112.
30. Sullivan, P. (2004). SUSTAINABLE MANAGEMENT OF SOIL-BORNE PLANT DISEASES. National Sustainable Agriculture Information Service, 1-16.
31. Tanmoy Das, M. Hajong, D. Majumdar, R. K. Tombisana Devi and T. Rajesh. (2016). CLIMATE CHANGE IMPACTS ON PLANT DISEASES. SAARC J. Agri, 14(2): 200-209 .
32. West et al. (2015). Impact of climate change on diseases in sustainable arable crop systems: CLIMDIS . Agriculture and horticulture development board, Project Report No. 539 .
33. Wiser, M. F. (2010). Protozoa and Human Disease. Ne York: Garland Science, 1-81.

34. Underexplored Players in Ecosystem Carbon Processing. American Society for Microbiology, 10.1128/mSystems.00076-18.

35. Graham Stirling, Helen Hayden, Tony Pattison, & Marcelle Stirling. (2016). Soil Health, Soil Biology, Soil-borne Diseases and Sustainable Agriculture: A Guide. Clayton: Csiro Publishing, pp.280.

36. Hoorman, J. J. (2011). The Role of Soil Bacteria. Agriculture and Natural Resources, 1-4.

37. HUSSAIN, B. (2015). Modernization in plant breeding approaches for improving biotic stress resistance in crop plants. Turkish Journal of Agriculture and Forestry, 515-525.

38. Idowu, J. (2014). Principles of Soil Health Management . Extension Plant Sciences Online Training Series, 1-53.

39. Jackie Aislabie & Julie R. Deslippe . (2013). SOIL MICROBES AND THEIR CONTRIBUTION TO SOIL SERVICES. SOIL MICROBIAL DIVERSITY, 143-159.

40. Jenkins, A. (2005). Soil fungi . soil biology basics, 1-2.

41. JohnClarkson,JaneDebode,LorenzoFurlan,RoyNeilson,AnnCharlotte Wallenhammar and Johann Zahrl. (Macromolecules: Indian Journal). Isolation and Identification of Fungi from Contaminated Soil to Build Biological Resource as Biocontrol Activity. SOIL-BORNE DISEASES, 1-4.

42. José M. De la Rosa, Ana Z. Miller & Heike Knicker . (2018). Soil-borne fungi challenge the concept of long-term biochemical recalcitrance of pyrochar. Scientifc Reports.

43. Katan, J. (2017). DISEASES CAUSED BY SOIL-BORNE PATHOGENS: BIOLOGY, MANAGEMENT AND CHALLENGES. Journal of Plant Pathology, 99 (2), 305-315.

44. Lwoff, A. (2014). Biochemistry and Physiology of Protozoa. New York: Elsevier, 1-27.

45. Magdalena Frąc, Silja E. Hannula, Marta Bełka, and Małgorzata Jędryczka. (2018). Fungal Biodiversity and Their Role in Soil Health. Frontier in microbiology, 9: 707.

46. Magdoff, F. (2001). Concept, Components, and Strategies of Soil Health in Agroecosystems. Journal of Nematology, 33(4):169–172.

47. Mark Schonbeck & Michael Stein. (2014). Reducing Risk through

Best Soil Health Management Practices in Organic Crop Production. Organic Farming Research Foundation, 1-54.

48. Niall A. Logan & Paul De Vos. (2011). Endospore-forming Soil Bacteria. Berlin: Springer Science & Business Media, 1-89.
49. Patel, D. A. (2016). ADDRESSING SOIL HEALTH MANAGEMENT ISSUES IN INDIA. International Journal of Research - GRANTHAALAYAH, Vol.4 (Iss.12) 1-14.
50. Peter J. Hotez, Donald A. P. Bundy, Kathleen Beegle, Simon Brooker, Lesley Drake, Nilanthi de Silva, Antonio Montresor, Dirk Engels, Matthew Jukes, Lester Chitsulo, Jeffrey Chow, Ramanan Laxminarayan, Catherine M. Michaud, Jeff Bethony, Rodrigo Correa-Oli. (2006). Helminth Infections: Soil Transmitted Helminth Infections and Schistosomiasis. Disease Control Priorities in Developing Countries, 467-481.
51. Puri, S. (2016). Eco-friendly management strategies for soil-borne plant pathogens. International Journal of Advanced Research in Biological Sciences, 3(1): 69-75.
52. R.N. Doetsch & T.M. Cook. (2012). Introduction to Bacteria and Their Ecobiology. Berlin: Springer Science & Business Media, 14-86.
53. Reddy, P. P. (2011). Biofumigation and Solarization for Management of Soil-Borne Plant Pathogens. London: Scientific Publishers, 1-93.
54. Sukhareva-Buell, N. (2003). Biologically Active Substances of Protozoa. Berlin: Springer Science & Business Media, pp.112.
55. Sullivan, P. (2004). SUSTAINABLE MANAGEMENT OF SOIL-BORNE PLANT DISEASES. National Sustainable Agriculture Information Service, 1-16.
56. Tanmoy Das, M. Hajong, D. Majumdar, R. K. Tombisana Devi and T. Rajesh. (2016). CLIMATE CHANGE IMPACTS ON PLANT DISEASES. SAARC J. Agri, 14(2): 200-209 .
57. West et al. (2015). Impact of climate change on diseases in sustainable arable crop systems: CLIMDIS . Agriculture and horticulture development board, Project Report No. 539 .
58. Wiser, M. F. (2010). Protozoa and Human Disease. Ne York: Garland Science, 1-81.

INDEX

M

N

O

P

T

V

W

Z